What does this painting have to do with math?

Swiss-born artist Paul Klee was interested in using color to express emotion. Here he created a grid, or array, of 35 colorful squares arranged in 5 rows and 7 columns. We will learn how an array helps us understand a larger shape by looking at the smaller shapes inside. Learning more about arrays will help us notice patterns and structure— an important skill for multiplication and division.

On the cover

Farbtafel "qu 1," 1930
Paul Klee, Swiss, 1879–1940
Pastel on paste paint on paper, mounted on cardboard
Kunstmuseum Basel, Basel, Switzerland

Paul Klee (1879–1940), *Farbtafel "qu 1"* (*Colour Table "Qu 1"*), 1930, 71. Pastel on coloured paste on paper on cardboard, 37.3 x 46.8 cm. Kunstmuseum Basel, Kupferstichkabinett, Schenkung der Klee-Gesellschaft, Bern. © 2020 Artists Rights Society (ARS), New York.

EUREKA
MATH²

GREAT
MINDS
™

Great Minds® is the creator of *Eureka Math*®,
Wit & Wisdom®, *Alexandria Plan*™, and *PhD Science*®.

Published by Great Minds PBC.
greatminds.org

Printed in the USA
B-Print

3 4 5 6 7 8 9 10 11 12 QDG 27 26 25 24 23

ISBN 978-1-64497-095-9

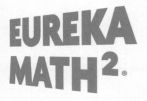

EUREKA
MATH².

A Story of Units®

Units of Any Number ▸ 3

LEARN

Contents

Multiplication and Division with Units of 0, 1, 6, 7, 8, and 9

Topic D
Multiplication with Multiples of 10 and Further Application of Concepts

Counting Collection 1

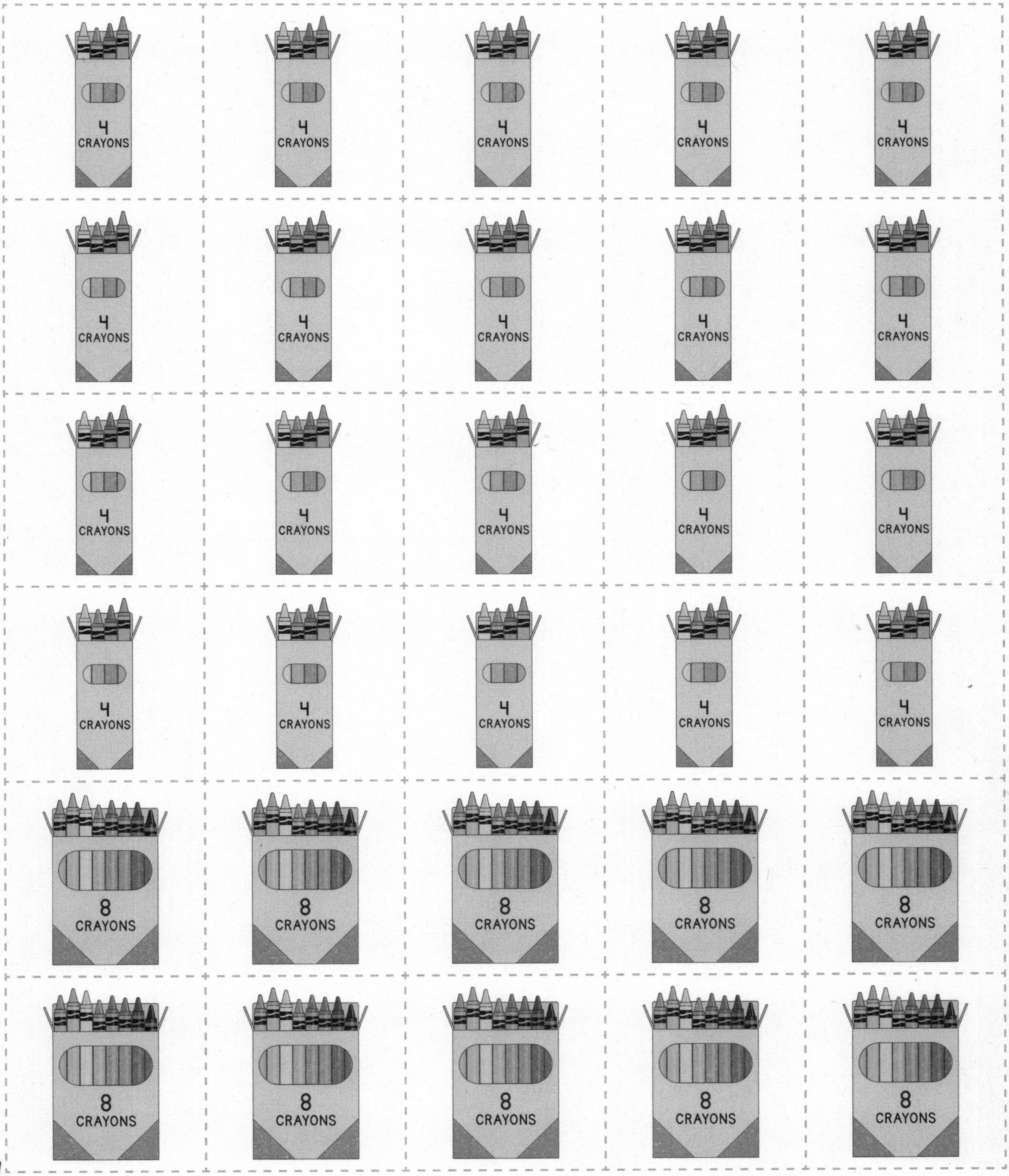

Counting Collection 2

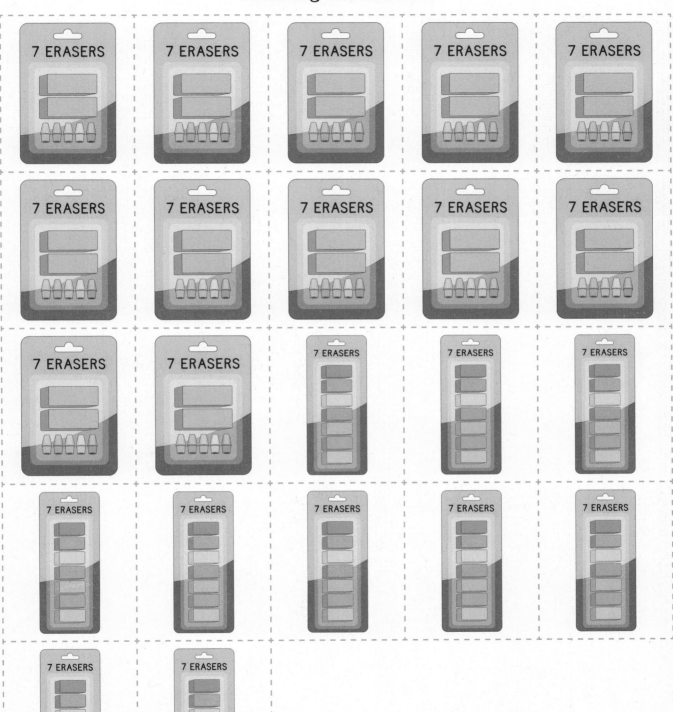

Counting Collection 3

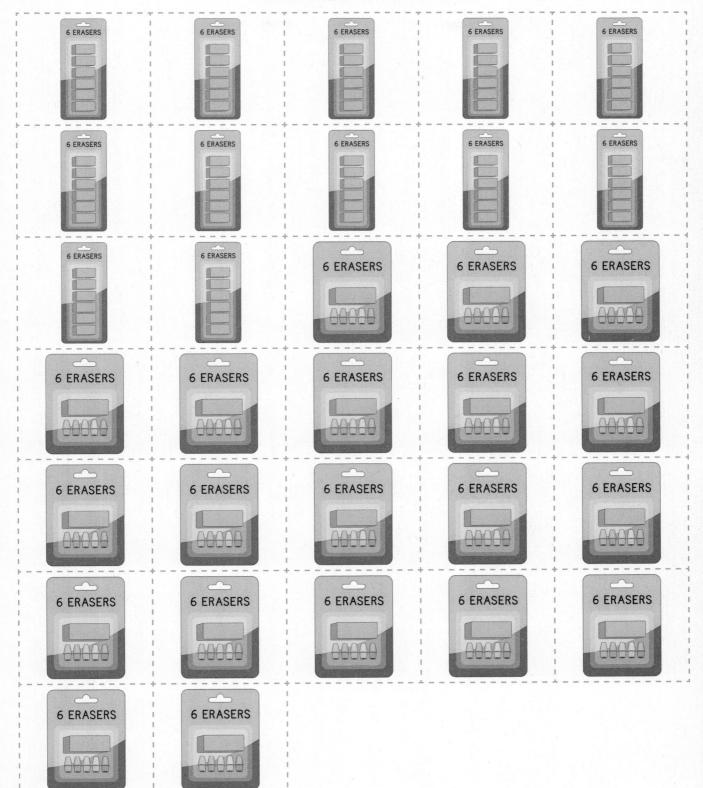

Counting Collection 4

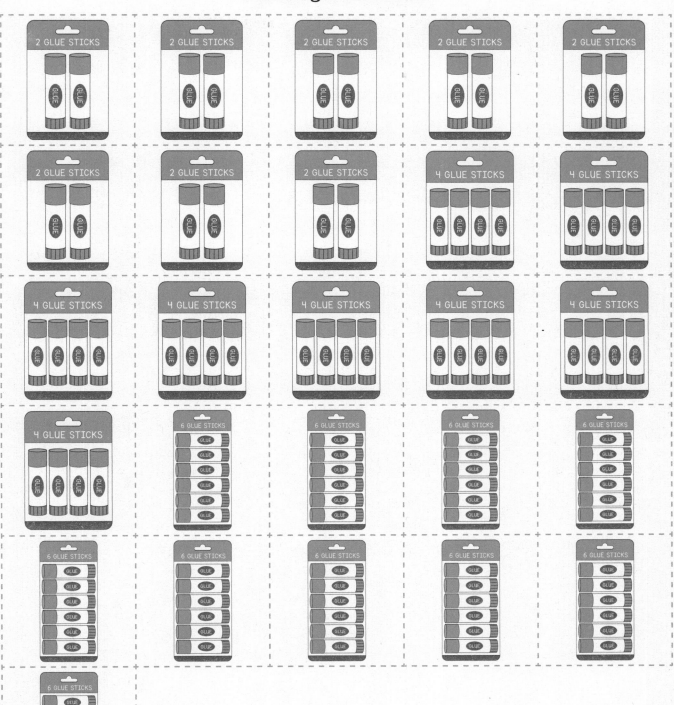

Counting Collection 5

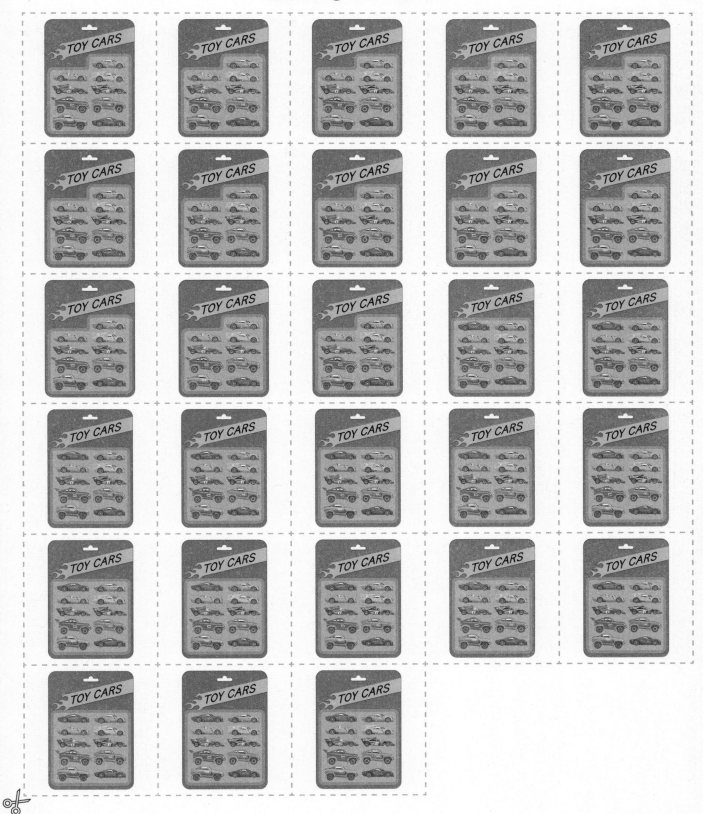

Oliver n rvi 75

Name

For this counting collection, I am partners with ___Oiver & Ava___.

We are counting ___Crayons___.

We estimate there are about ___80___ of them.

This is how we organized and counted the collection:

16 is each
colum
16 x 5 = 80

We counted _____ altogether.

An equation that describes how we found the total is:

_____

Self-Reflection

Write one thing that worked well for you and your partner. Explain why it worked well.

_____

_____

_____

Write one challenge you had. How did you work through the challenge?

_____

_____

_____

Name

Use the Read–Draw–Write process to solve the problem.

1. There are 48 teacher mailboxes in the office. The mailboxes are in 6 equal rows.

   a. How many mailboxes are in each row?

   There are 8 mail boxes in each row. 48 $48 \div 6 = 8$

   | 8 | 8 | 8 | 8 | 8 | 8 |

   b. Write an unknown factor equation that represents the problem.

   $6 \cdot m = 48$

Find the value of each unknown.

2. $7 \times 6 = k$ $k = \underline{42}$

3. $48 \div 6 = r$ $r = \underline{8}$

4. $6 \times p = 36$ $p = \underline{6}$

5. $54 \div h = 6$ $h = \underline{9}$

6. $w \div 3 = 6$ $w = \underline{18}$

2

Name

1. Each bunch of balloons has 3 red balloons and 3 purple balloons.

   a. Skip-count by threes to find the total number of balloons.

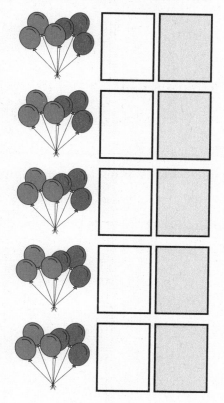

   b. Complete the statements.

   10 threes is 30 .

   10 × 3 = 30

   5 sixes is 30 .

   5 × 6 = 30

   c. Use the pictures of balloons to help you complete the statement.

   2 groups of 5 × _____ is the same as 5 × _____ .

Find the value of the unknown.

2. $6 \times 2 = c$

    $c = \underline{12}$

3. $6 \times w = 12$

    $w = \underline{2}$

4. $f \times 6 = 18$

    $f = \underline{3}$

5. $18 \div 6 = a$

    $a = \underline{3}$

6. $30 \div d = 5$

    $d = \underline{6}$

7. $h \div 3 = 6$

    $h = \underline{18}$

8. $6 \times k = 54$

    $k = \underline{9}$

9. $60 \div 6 = m$

    $m = \underline{10}$

10. $p \times 6 = 66$

    $p = \underline{11}$

11. Mia puts a total of 42 marbles into 6 boxes. Each box has an equal number of marbles. How many marbles are in each box?

    a. Draw and label a tape diagram that represents the problem. Label the unknown as m.

    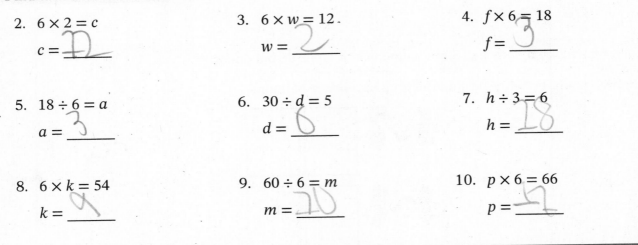

    b. Write a division equation to represent the problem. Use the letter m for the unknown. Then find the value of m.

    $42 \div 6 = m$

    $m = 7$

12. Mr. Davis plants tulip bulbs in his garden. He plants 3 rows with 6 bulbs in each row. How many tulip bulbs does Mr. Davis plant?

a. Does the tape diagram below represent the problem? Explain.

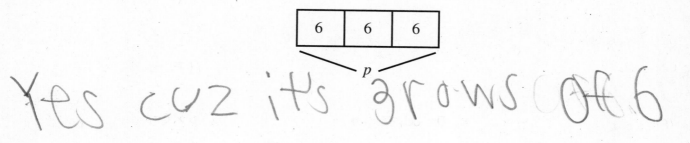

Yes cuz its grows of 6

b. Write an equation to represent the problem. Use the letter p for the unknown. Then find the value of p.

$3 \times 6 = p$
$p = 8$

Name _____

1. Color 10 fours blue and 10 fours orange.

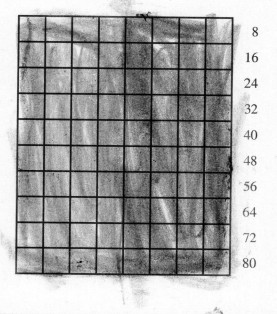

8
16
24
32
40
48
56
64
72
80

Find the value of each unknown.

2. $c \times 8 = 24$ $c = \underline{3}$

3. $40 \div m = 8$ $m = \underline{5}$

4. $8 \times 4 = h$ $h = \underline{32}$

5. $w \div 8 = 8$ $w = \underline{64}$

3

Name _____

1. Complete parts (a)–(c) to show the relationship between fours and eights.

   a. Skip-count by fours.

   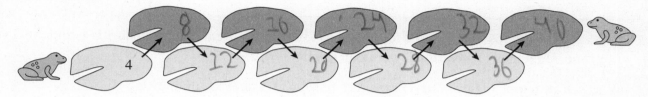

   b. Complete each statement to find the total.

   10 fours is __4__.

   _10_ × 4 = _40_

   5 eights is __4__.

   _5_ × 8 = _40_

   c. Complete the statement to show the connection between the fours and eights.

   2 groups of 5 × __4__ is the same as 5 × __8__.

Find the value of each unknown.

2. $4 \times 2 = y$

   $y = \underline{8}$

3. $8 \times w = 8$

   $w = \underline{1}$

4. $k \times 2 = 16$

   $k = \underline{8}$

5. $16 \div 8 = z$

   $z = \underline{2}$

6. $p \div 8 = 6$

   $p = \underline{48}$

7. $40 \div r = 8$

   $r = \underline{5}$

8. $56 \div 8 = v$

   $v = \underline{7}$

9. $6 \times 8 = n$

   $n = \underline{48}$

10. $m \div 8 = 8$

    $m = \underline{64}$

11. Write an equation to represent the tape diagram below. Then find the value of the unknown.

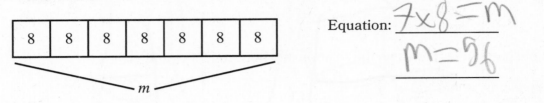

Equation: $7 \times 8 = m$

$m = 56$

12. Mr. Lopez groups his students for a field trip. There is a total of 8 groups with 4 students in each group. How many students are going on the field trip?

a. Draw a tape diagram to represent the problem. Label the unknown as c.

b. Write an equation using c to represent the total number of students. Then find the value of c.

$8 \times 4 = c$ $c = 32$

13. Luke bakes 72 cookies for a bake sale. He divides them equally among 8 plates. How many cookies are on each plate?

a. Draw a tape diagram to represent the problem. Label the unknown as y.

b. Write a division equation using y to represent the unknown. Then find the value of y.

c. A classmate says this problem can also be represented using the equation $8 \times y = 72$. Is the classmate correct? Why?

Name _____

1. Use the array to help you fill in the blanks.

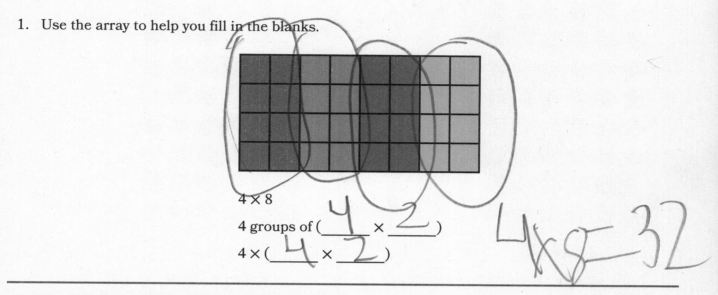

4×8

4 groups of (__4__ × __2__)

$4 \times ($ __4__ × __2__ $)$

$4 \times 8 = 32$

2. Use the array to help you fill in the blanks.

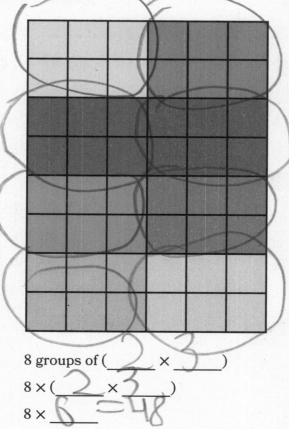

8 groups of (__2__ × __3__)

$8 \times ($ __2__ × __3__ $)$

$8 \times$ __6__ $= 48$

Circle equal groups in each array and then use the array to help you fill in the blanks.

3.

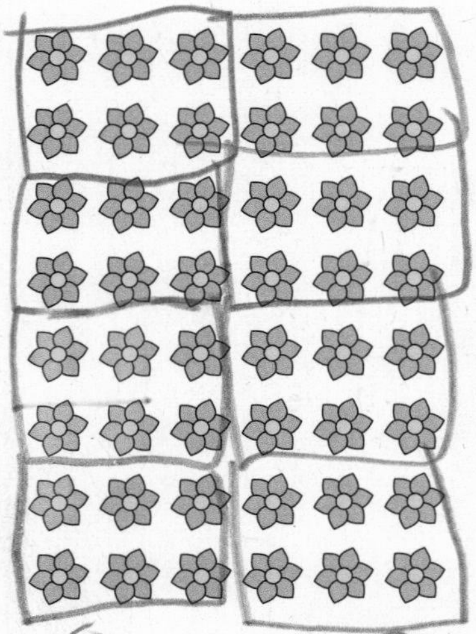

__8__ groups of (__2__ × __3__)

__8__ × (__2__ × __3__)

__8__ × __6__ = 48

4.

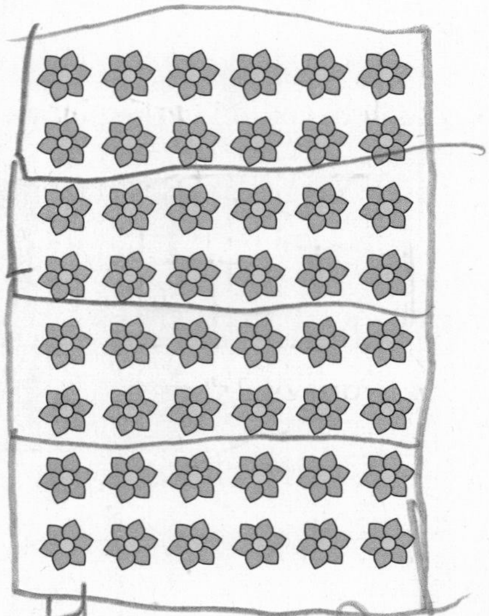

__4__ groups of (__2__ × __6__)

__4__ × (__2__ × __6__)

__4__ × __12__ = 48

5.

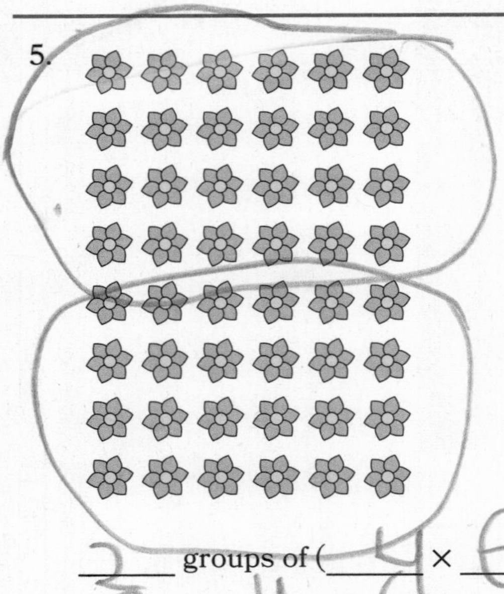

__2__ groups of (__4__ × __6__)

__2__ × (__4__ × __6__)

__2__ × __24__ = 48

6.

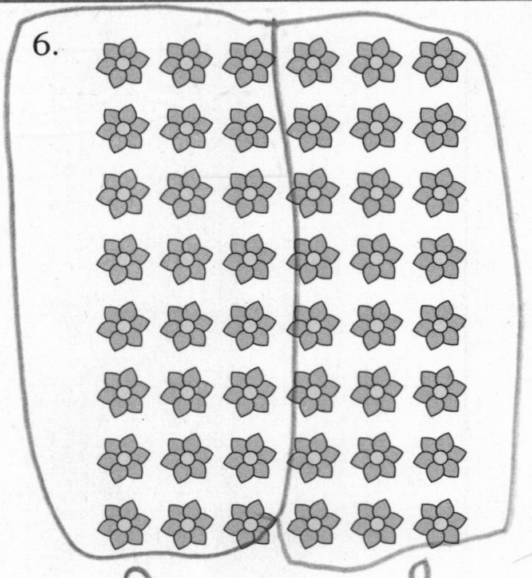

__2__ groups of (__8__ × __3__)

__2__ × (__8__ × __3__)

__2__ × __24__ = 48

Name _____

1. Use the arrays to help you fill in the blanks.

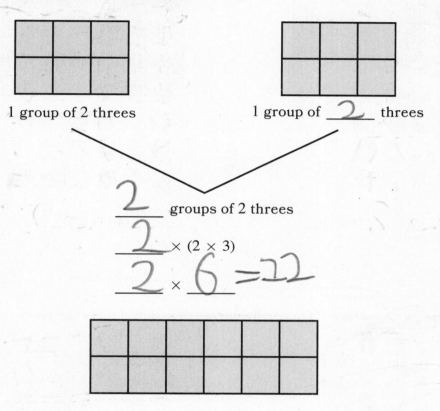

1 group of 2 threes   1 group of __2__ threes

__2__ groups of 2 threes

__2__ × (2 × 3)

__2__ × __6__ = 12

Use the arrays to help you fill in the blanks.

2.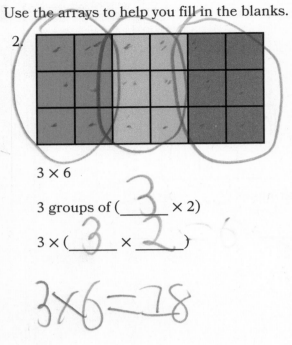

3 × 6

3 groups of (__3__ × 2)

3 × (__3__ × __2__) = 6

3 × 6 = 18

3.

6 × 4

6 groups of (__2__ × __2__)

6 × (__2__ × __2__) = 4

6 × 4 = 24

Circle equal groups in each array. Then use the arrays to help you fill in the blanks.

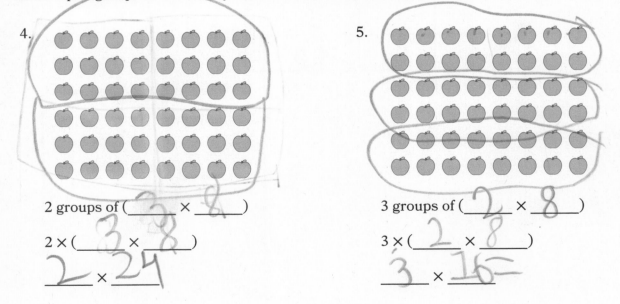

4.

2 groups of (__3__ × __8__)

2 × (__3__ × __8__)

__2__ × __24__

5.

3 groups of (__2__ × __8__)

3 × (__2__ × __8__)

__3__ × __16__

6.

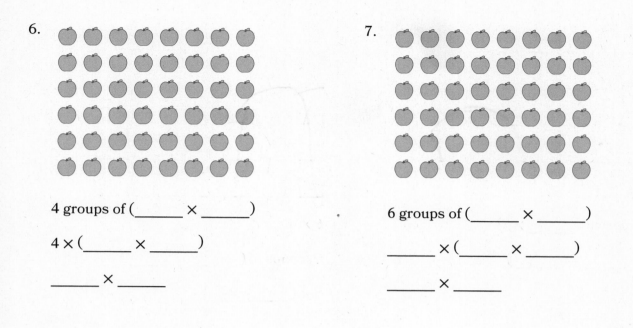

7.

4 groups of (_____ × _____)

4 × (_____ × _____)

_____ × _____

6 groups of (_____ × _____)

_____ × (_____ × _____)

_____ × _____

8. Draw an array to show 4×6. Then circle equal groups in your array to show $4 \times 6 = 4 \times (2 \times 3)$.

Sprint

Round to the nearest ten.

1.	54 ≈ _____	
2.	138 ≈ _____	

A

Number Correct: _____

Round to the nearest ten.

1.	29 ≈ _____	
2.	49 ≈ _____	
3.	89 ≈ _____	
4.	34 ≈ _____	
5.	64 ≈ _____	
6.	94 ≈ _____	
7.	36 ≈ _____	
8.	63 ≈ _____	
9.	58 ≈ _____	
10.	85 ≈ _____	
11.	99 ≈ _____	
12.	241 ≈ _____	
13.	246 ≈ _____	
14.	419 ≈ _____	
15.	412 ≈ _____	
16.	647 ≈ _____	
17.	641 ≈ _____	
18.	853 ≈ _____	
19.	858 ≈ _____	
20.	924 ≈ _____	
21.	926 ≈ _____	
22.	928 ≈ _____	

23.	337 ≈ _____	
24.	307 ≈ _____	
25.	563 ≈ _____	
26.	503 ≈ _____	
27.	766 ≈ _____	
28.	706 ≈ _____	
29.	894 ≈ _____	
30.	804 ≈ _____	
31.	932 ≈ _____	
32.	902 ≈ _____	
33.	361 ≈ _____	
34.	555 ≈ _____	
35.	505 ≈ _____	
36.	497 ≈ _____	
37.	507 ≈ _____	
38.	698 ≈ _____	
39.	708 ≈ _____	
40.	996 ≈ _____	
41.	1,654 ≈ _____	
42.	1.057 ≈ _____	
43.	1,606 ≈ _____	
44.	1,008 ≈ _____	

B

Number Correct: _____

Improvement: _____

Round to the nearest ten.

1.	19 ≈ _____	
2.	39 ≈ _____	
3.	79 ≈ _____	
4.	24 ≈ _____	
5.	54 ≈ _____	
6.	84 ≈ _____	
7.	26 ≈ _____	
8.	62 ≈ _____	
9.	57 ≈ _____	
10.	75 ≈ _____	
11.	99 ≈ _____	
12.	141 ≈ _____	
13.	146 ≈ _____	
14.	319 ≈ _____	
15.	312 ≈ _____	
16.	547 ≈ _____	
17.	541 ≈ _____	
18.	753 ≈ _____	
19.	758 ≈ _____	
20.	824 ≈ _____	
21.	826 ≈ _____	
22.	828 ≈ _____	

23.	278 ≈ _____	
24.	208 ≈ _____	
25.	464 ≈ _____	
26.	404 ≈ _____	
27.	657 ≈ _____	
28.	607 ≈ _____	
29.	792 ≈ _____	
30.	702 ≈ _____	
31.	843 ≈ _____	
32.	803 ≈ _____	
33.	261 ≈ _____	
34.	555 ≈ _____	
35.	505 ≈ _____	
36.	398 ≈ _____	
37.	408 ≈ _____	
38.	599 ≈ _____	
39.	609 ≈ _____	
40.	997 ≈ _____	
41.	1,653 ≈ _____	
42.	1,058 ≈ _____	
43.	1,607 ≈ _____	
44.	1,009 ≈ _____	

5

Name

1. Liz sees a display of cans of peas at the grocery store. How many cans of peas are in the display?

   Use the break apart and distribute strategy to find the total number of cans of peas.

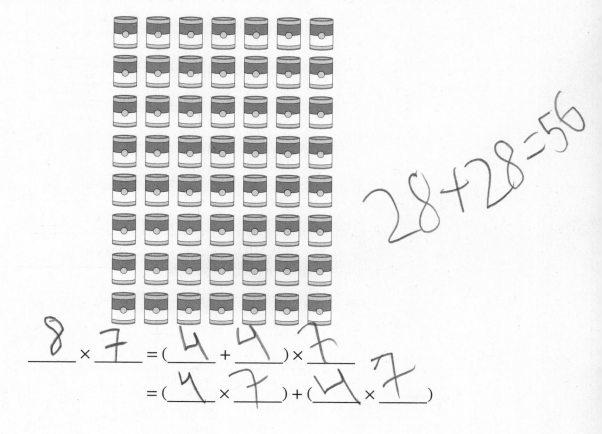

$$\underline{8} \times \underline{7} = (\underline{4} + \underline{4}) \times \underline{7}$$

$$28 + 28 = 56$$

$$= (\underline{4} \times \underline{7}) + (\underline{4} \times \underline{7})$$

2. Liz sees a display of sauce at the grocery store. How many jars of sauce are in the display?

    Use the break apart and distribute strategy to find the total number of jars of sauce.

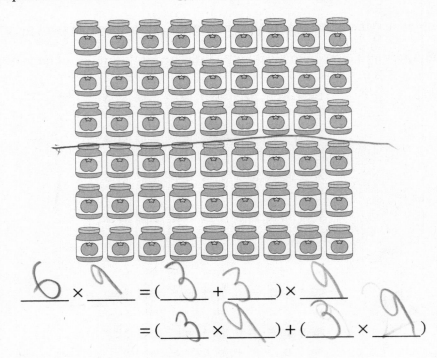

$$\underline{6} \times \underline{9} = (\underline{3} + \underline{3}) \times \underline{9}$$

$$= (\underline{3} \times \underline{9}) + (\underline{3} \times \underline{9})$$

3. Liz sees a display of 8 rows of tomatoes at the grocery store. There are 9 tomatoes in each row. How many tomatoes are in the display?

a. Draw a tape diagram to represent the tomatoes.

b. Use the break apart and distribute strategy to find the total number of tomatoes.

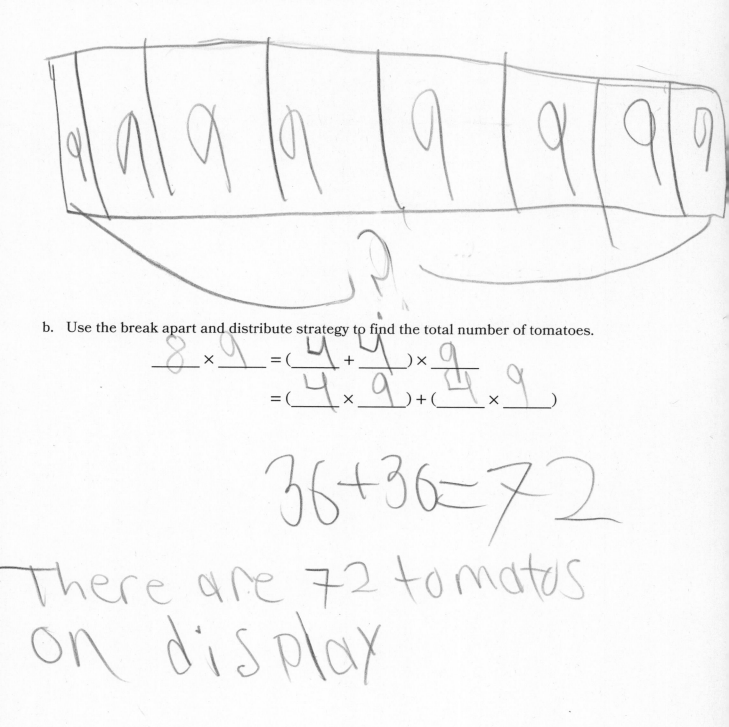

$$\underline{8} \times \underline{9} = (\underline{4} + \underline{4}) \times \underline{9}$$
$$= (\underline{4} \times \underline{9}) + (\underline{4} \times \underline{9})$$

$$36 + 36 = 72$$

There are 72 tomatos on display

Name

1. Draw part of the array to complete the number bond. Then use it to help you fill in the blanks and find the total.

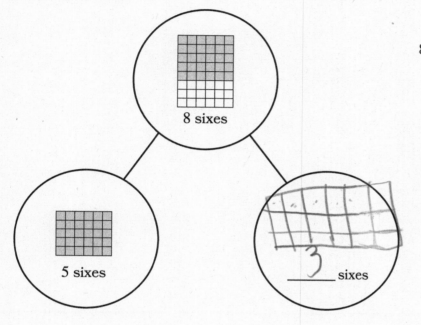

8 sixes = 5 sixes + $\underline{3}$ sixes

$8 \times 6 = (5 \times 6) + (\underline{3} \times 6)$

$8 \times 6 = \underline{30} + (\underline{18})$

$8 \times 6 = \underline{48}$

Use the arrays to help you fill in the blanks and find the totals.

2.

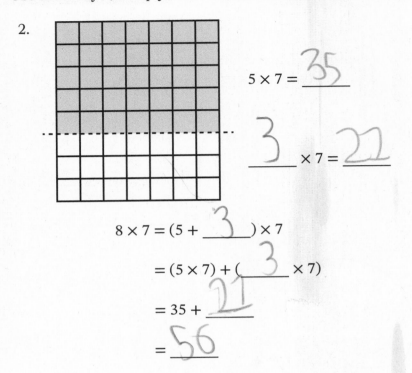

$5 \times 7 = \underline{35}$

$\underline{3} \times 7 = \underline{21}$

$8 \times 7 = (5 + \underline{3}) \times 7$

$= (5 \times 7) + (\underline{3} \times 7)$

$= 35 + \underline{21}$

$= \underline{56}$

3.

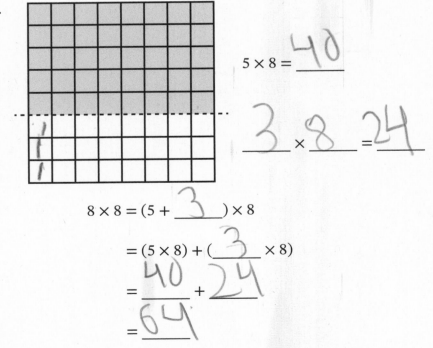

$5 \times 8 = \underline{40}$

$\underline{3} \times \underline{8} = \underline{24}$

$8 \times 8 = (5 + \underline{3}) \times 8$

$= (5 \times 8) + (\underline{3} \times 8)$

$= \underline{40} + \underline{24}$

$= \underline{64}$

Label the tape diagrams. Then complete the equations.

4. $(5 \times 6) =$ ___30___ $(\underline{2} \times 6) =$ ___12___

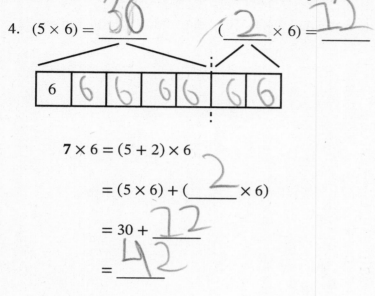

$7 \times 6 = (5 + 2) \times 6$

$= (5 \times 6) + (\underline{2} \times 6)$

$= 30 + \underline{12}$

$= \underline{42}$

5. $(\underline{5} \times 6) =$ ___30___ $(\underline{4} \times 6) =$ ___24___

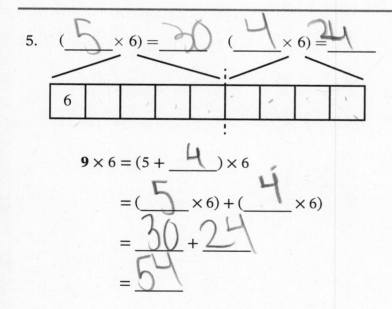

$9 \times 6 = (5 + \underline{4}) \times 6$

$= (\underline{5} \times 6) + (\underline{4} \times 6)$

$= \underline{30} + \underline{24}$

$= \underline{54}$

6. Adam walks 9 laps around the track every day for 8 days. How many total laps does he walk?

    a. To find the total, Jayla breaks 8×9 into 5×9 and 3×9. Then she adds 45 and 27 to get 72. Explain why her strategy works.

    b. Show another way 8×9 can be broken apart into smaller facts to find the product.

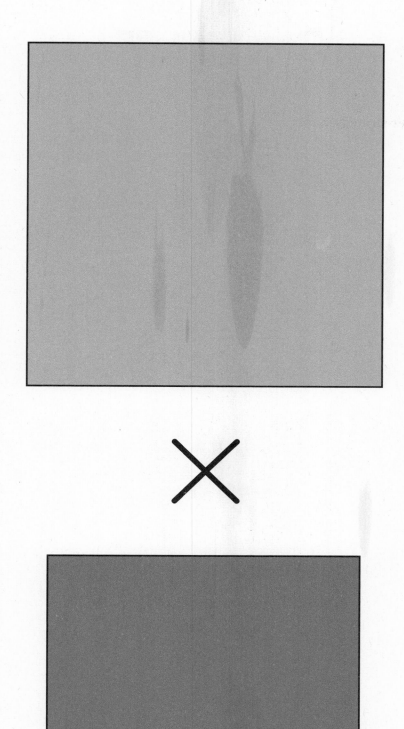

Name _____

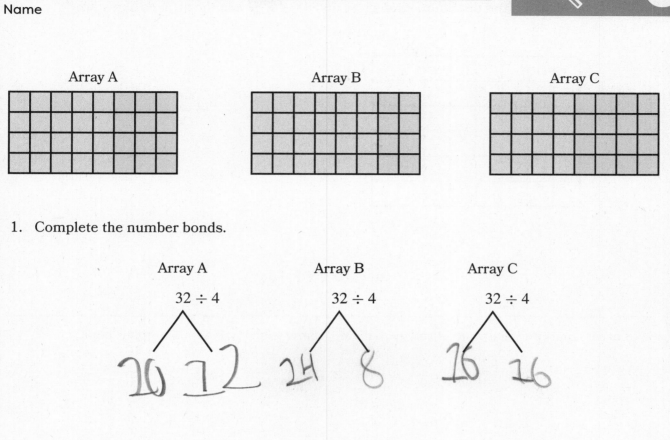

Array A

Array B

Array C

1. Complete the number bonds.

Array A

$32 \div 4$

20   12

Array B

$32 \div 4$

24   8

Array C

$32 \div 4$

16   16

2. Shade the array and complete the number bond to show a different way to break apart 32.

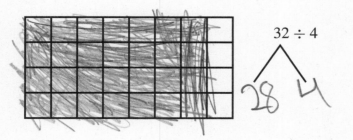

$32 \div 4$

28   4

3. Use the break apart and distribute strategy to find 48 ÷ 6. Shade the array to show how to break apart 48.

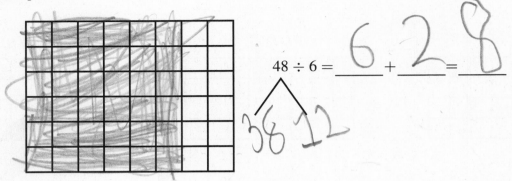

$48 \div 6 = \underline{6} + \underline{2} = \underline{8}$

36   12

Use the break apart and distribute strategy to divide. Show your work with a number bond.

4. $28 \div 4$

5. $54 \div 6$

6. $48 \div 8$

Name _____

Use the array to help you complete each number bond and equation. The first one is started for you.

1.

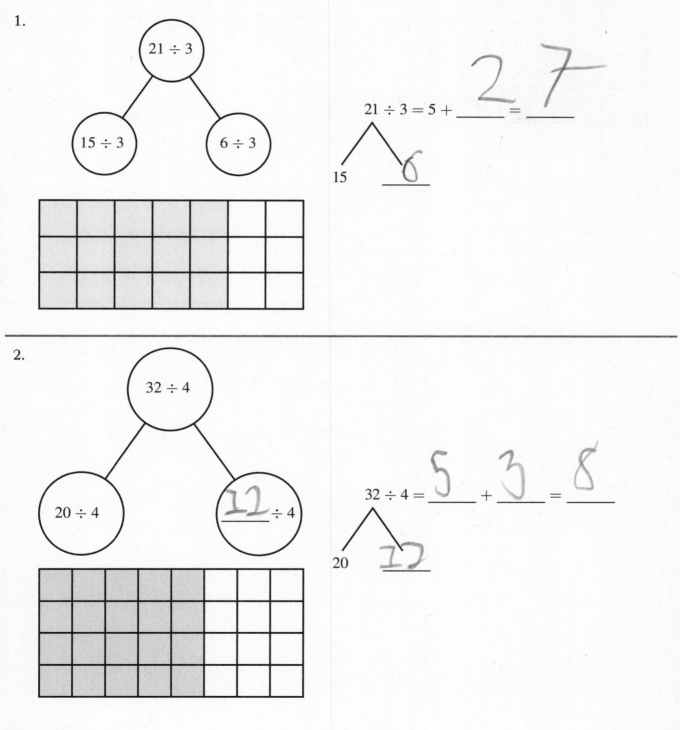

$21 \div 3 = 5 + \underline{2} = \underline{7}$

15 _____ 6

2.

$32 \div 4 = \underline{5} + \underline{3} = \underline{8}$

20 _____ 12

Use the array to help you complete each equation.

3.

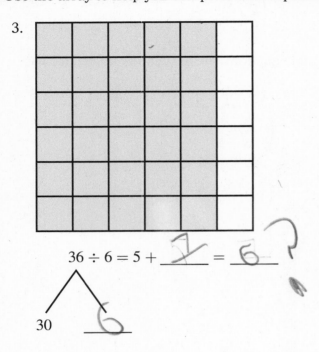

$36 \div 6 = 5 + \underline{7} = \underline{6}$

30   6

4.

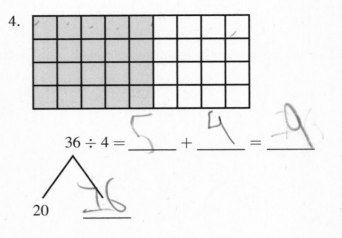

$36 \div 4 = \underline{5} + \underline{4} = \underline{9}$

20   16

Use the break apart and distribute strategy to divide.

5. $42 \div 6 = 5 + \underline{2} = \underline{7}$

30   12

6. $56 \div 8 = \underline{5} + \underline{2} = \underline{7}$

40   16

7. $48 \div 8 = \underline{5} + \underline{2} = \underline{6}$

40   8

8. $54 \div 6 = \underline{5} + \underline{4} = \underline{9}$

30   24

9. Use the break apart and distribute strategy to find $64 \div 8$. Explain your thinking.

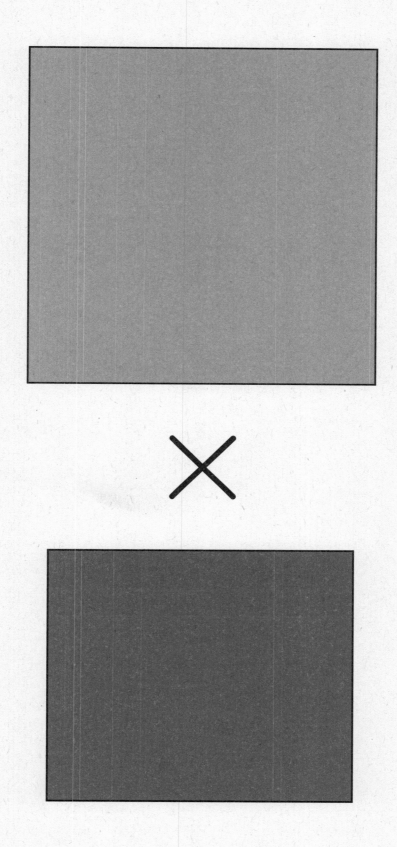

Name

1. David has basketball practice every Monday starting in January. Complete the table to show what dates in January he has practice.

Practice	1	2	3	4
Date in January	7			

JANUARY						
S	M	T	W	T	F	S
		1	2	3	4	5
6	7	8	9	10	11	12
13	14	15	16	17	18	19
20	21	22	23	24	25	26
27	28	29	30	31		

2. Gabe plants his flowers in rows. How many flowers does Gabe plant?

	Unit Form	Multiplication Equation	Division Equation

3. Gabe has 56 seeds that he plants in rows of 7. How many rows of seeds does Gabe plant?

4. Deepa has 42 oranges. She puts all the oranges in bags. There are 7 oranges in each bag. How many bags does she use?

---

5. Adam packs 49 oranges into 7 bags.

   a. How many oranges should he pack in each bag so that the bags all have the same number?

   b. Adam gives 3 bags of oranges to his sister. How many oranges does Adam give his sister?

Name _____

1. Use the array to skip-count by sevens. Then complete each equation.

$1 \times 7 = 7$	$7 \div 7 =$ _____
$2 \times 7 =$ _____	$14 \div 7 =$ _____
$3 \times 7 =$ _____	_____ $\div 7 =$ _____
$4 \times 7 =$ _____	_____ $\div 7 =$ _____
$5 \times 7 =$ _____	_____ $\div 7 =$ _____
$6 \times 7 =$ _____	_____ $\div 7 =$ _____
$7 \times 7 =$ _____	_____ $\div 7 =$ _____
$8 \times 7 =$ _____	_____ $\div 7 =$ _____
$9 \times 7 =$ _____	_____ $\div 7 =$ _____
$10 \times 7 =$ _____	_____ $\div 7 =$ _____

2. Robin puts apples in bags. Each bag has 7 apples. The table shows how many apples are needed for different numbers of bags. Complete the table.

Number of Bags	1	2		4	5		7	8	9
Total Number of Apples	7		21	28		42	49		

Find the value of each unknown.

3. $7 \times 2 = p$

   $p =$ _____

4. $2 \times f = 14$

   $f =$ _____

5. $y \times 7 = 28$

   $y =$ _____

6. $21 \div 7 = m$

   $m =$ _____

7. $35 \div a = 5$

   $a =$ _____

8. $28 \div c = 7$

   $c =$ _____

9. $g \times 7 = 56$

   $g =$ _____

10. $7 \times r = 63$

    $r =$ _____

11. $d \div 7 = 7$

    $d =$ _____

12. Write an equation to represent the tape diagram.

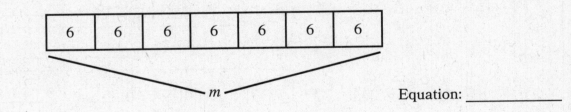

Equation: _____

PROBLEM SET

Use the Read–Draw–Write process to solve each problem.

13. Pablo's bookcase has 4 shelves. There are 7 books on each shelf. How many books are in the bookcase?

14. Carla collects 35 acorns. She divides them equally into 7 piles. How many acorns are in each pile?

Name

Use the Read–Draw–Write process to solve the problem.

Gabe puts 63 apples into 7 boxes. Each box has the same number of apples. How many apples does Gabe put in each box?

Sprint

Round to the nearest hundred.

1.	379 ≈ _____	
2.	309 ≈ _____	

A

Number Correct: _____

Round to the nearest hundred.

1.	299 ≈ _____	
2.	499 ≈ _____	
3.	899 ≈ _____	
4.	344 ≈ _____	
5.	644 ≈ _____	
6.	944 ≈ _____	
7.	316 ≈ _____	
8.	516 ≈ _____	
9.	716 ≈ _____	
10.	855 ≈ _____	
11.	655 ≈ _____	
12.	355 ≈ _____	
13.	496 ≈ _____	
14.	476 ≈ _____	
15.	426 ≈ _____	
16.	788 ≈ _____	
17.	768 ≈ _____	
18.	748 ≈ _____	
19.	832 ≈ _____	
20.	852 ≈ _____	
21.	872 ≈ _____	
22.	842 ≈ _____	

23.	374 ≈ _____	
24.	370 ≈ _____	
25.	304 ≈ _____	
26.	538 ≈ _____	
27.	530 ≈ _____	
28.	508 ≈ _____	
29.	263 ≈ _____	
30.	260 ≈ _____	
31.	203 ≈ _____	
32.	555 ≈ _____	
33.	550 ≈ _____	
34.	505 ≈ _____	
35.	299 ≈ _____	
36.	909 ≈ _____	
37.	999 ≈ _____	
38.	990 ≈ _____	
39.	1,485 ≈ _____	
40.	1,480 ≈ _____	
41.	1,405 ≈ _____	
42.	1,085 ≈ _____	
43.	1,080 ≈ _____	
44.	1,005 ≈ _____	

B

Number Correct: _____

Improvement: _____

Round to the nearest hundred.

1.	199 ≈ _____	
2.	399 ≈ _____	
3.	799 ≈ _____	
4.	244 ≈ _____	
5.	544 ≈ _____	
6.	844 ≈ _____	
7.	216 ≈ _____	
8.	416 ≈ _____	
9.	616 ≈ _____	
10.	755 ≈ _____	
11.	455 ≈ _____	
12.	255 ≈ _____	
13.	396 ≈ _____	
14.	376 ≈ _____	
15.	326 ≈ _____	
16.	688 ≈ _____	
17.	668 ≈ _____	
18.	648 ≈ _____	
19.	732 ≈ _____	
20.	752 ≈ _____	
21.	772 ≈ _____	
22.	742 ≈ _____	

23.	274 ≈ _____	
24.	270 ≈ _____	
25.	204 ≈ _____	
26.	438 ≈ _____	
27.	430 ≈ _____	
28.	408 ≈ _____	
29.	163 ≈ _____	
30.	160 ≈ _____	
31.	103 ≈ _____	
32.	555 ≈ _____	
33.	550 ≈ _____	
34.	505 ≈ _____	
35.	199 ≈ _____	
36.	908 ≈ _____	
37.	988 ≈ _____	
38.	980 ≈ _____	
39.	1,475 ≈ _____	
40.	1,470 ≈ _____	
41.	1,405 ≈ _____	
42.	1,075 ≈ _____	
43.	1,070 ≈ _____	
44.	1,005 ≈ _____	

Name _____

1. The popcorn boxes at the snack bar are arranged in 8 rows. There are 7 boxes in each row. How many boxes of popcorn are at the snack bar?

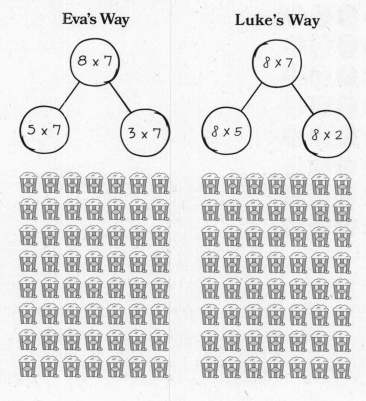

2. A carnival game has a wall of balloons. How many balloons are on the wall?

a. Draw a line to break apart the columns in the array.

b. Use the break apart and distribute strategy to multiply. Then write a solution statement.

$$9 \times 7 = \underline{\hspace{1cm}} \times (\underline{\hspace{1cm}} + \underline{\hspace{1cm}})$$

3. There are 6 benches at the animal show. 7 people can sit on each bench. What is the total number of people that can sit on the benches?

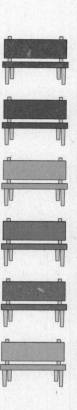

a. Draw an array to represent the problem. Then draw a line to break apart the array.

b. Use the break apart and distribute strategy to multiply. Then write a solution statement.

Name _____

Find 4×7 by using the break apart and distribute strategy in two different ways. A line is drawn in each array to help you.

1.

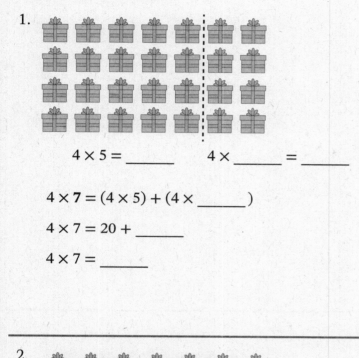

$4 \times 5 = $ _____ $4 \times$ _____ $= $ _____

$4 \times \mathbf{7} = (4 \times 5) + (4 \times$ _____ $)$

$4 \times 7 = 20 + $ _____

$4 \times 7 = $ _____

2.

_____ $\times 7 = $ _____

_____ $\times 7 = $ _____

$\mathbf{4} \times 7 = (2 \times 7) + ($ _____ $\times 7)$

$4 \times 7 = 14 + $ _____

$4 \times 7 = $ _____

Find 6×7 by using the break apart and distribute strategy in two different ways. A line is drawn in each array to help you.

3.

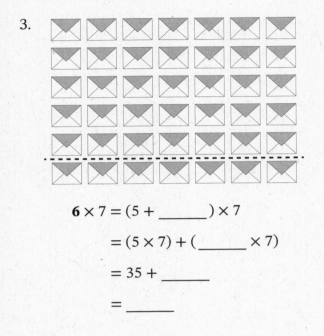

$$\mathbf{6} \times 7 = (5 + \underline{\hspace{1cm}}) \times 7$$
$$= (5 \times 7) + (\underline{\hspace{1cm}} \times 7)$$
$$= 35 + \underline{\hspace{1cm}}$$
$$= \underline{\hspace{1cm}}$$

4.

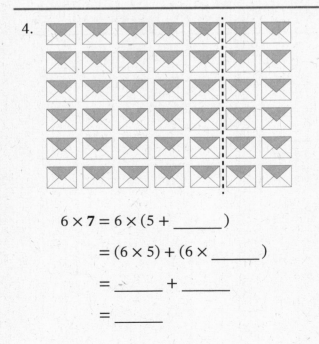

$$6 \times \mathbf{7} = 6 \times (5 + \underline{\hspace{1cm}})$$
$$= (6 \times 5) + (6 \times \underline{\hspace{1cm}})$$
$$= \underline{\hspace{1cm}} + \underline{\hspace{1cm}}$$
$$= \underline{\hspace{1cm}}$$

PROBLEM SET

Find 8 × 7 by using the break apart and distribute strategy in two different ways. Draw a line in each array to break it apart.

5.

6.

$8 \times 7 = (\underline{\hspace{1cm}} + \underline{\hspace{1cm}}) \times 7$

$= (\underline{\hspace{1cm}} \times 7) + (\underline{\hspace{1cm}} \times 7)$

$= \underline{\hspace{1cm}} + \underline{\hspace{1cm}}$

$= \underline{\hspace{1cm}}$

$8 \times 7 = 8 \times (\underline{\hspace{1cm}} + \underline{\hspace{1cm}})$

$= (8 \times \underline{\hspace{1cm}}) + (8 \times \underline{\hspace{1cm}})$

$= \underline{\hspace{1cm}} + \underline{\hspace{1cm}}$

$= \underline{\hspace{1cm}}$

Find 9×7 by using the break apart and distribute strategy in two different ways. Complete each number bond to help you.

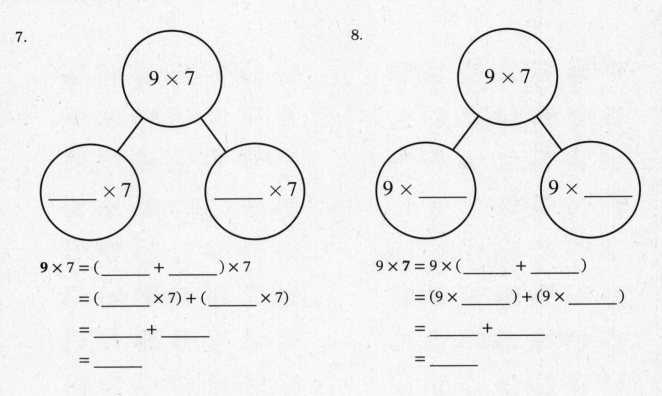

7.

$9 \times 7 = ($ _____ $+$ _____ $) \times 7$

$= ($ _____ $\times 7) + ($ _____ $\times 7)$

$=$ _____ $+$ _____

$=$ _____

8.

$9 \times 7 = 9 \times ($ _____ $+$ _____ $)$

$= (9 \times$ _____ $) + (9 \times$ _____ $)$

$=$ _____ $+$ _____

$=$ _____

Use the Read–Draw–Write process to solve each problem.

9. David organizes his coins into rows to make them easier to count. There are 7 rows with 7 coins in each row.

   How many coins does David have?

10. At the day care center, Casey sorts toys into bins. There are 8 bins, and she places 7 toys into each bin.

    How many toys does Casey sort?

Name

A band has 6 rows of trumpet players. Each row has 7 trumpet players. How many trumpet players are in the band?

Use the array to help you complete the equations.

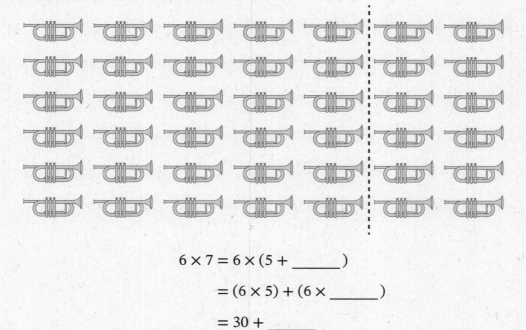

$$6 \times 7 = 6 \times (5 + \underline{\hspace{1.5cm}})$$

$$= (6 \times 5) + (6 \times \underline{\hspace{1.5cm}})$$

$$= 30 + \underline{\hspace{1.5cm}}$$

$$= \underline{\hspace{1.5cm}}$$

There are _____ trumpet players in the band.

Name _____

Rewrite each expression as a two-factor expression. Then find the products.

1. $(2 \times 4) \times 3$

2. $2 \times (4 \times 3)$

3. $(3 \times 2) \times 5$

4. $3 \times (2 \times 5)$

5. $(2 \times 1) \times 5$

6. $2 \times (1 \times 5)$

7. $(4 \times 2) \times 2$

8. $4 \times (2 \times 2)$

9. $16 \times 3 = 8 \times (2 \times 3)$

$16 \times 3 =$ _____ $\times$ _____

10. $16 \times 3 = (\underline{\hspace{1cm}} \times \underline{\hspace{1cm}}) \times 3$

$16 \times 3 = \underline{\hspace{1cm}} \times (\underline{\hspace{1cm}} \times \underline{\hspace{1cm}})$

$16 \times 3 = \underline{\hspace{1cm}} \times \underline{\hspace{1cm}}$

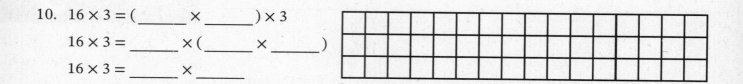

11. $15 \times 3 = (\underline{\hspace{1cm}} \times \underline{\hspace{1cm}}) \times \underline{\hspace{1cm}}$

$15 \times 3 = \underline{\hspace{1cm}} \times (\underline{\hspace{1cm}} \times \underline{\hspace{1cm}})$

$15 \times 3 = \underline{\hspace{1cm}} \times \underline{\hspace{1cm}}$

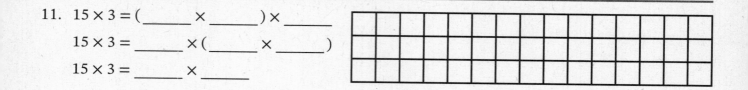

Name _____

Circle to show the equal groups in each array. Then circle the expression that represents the equal groups.

1. 3 groups of 2 × 4

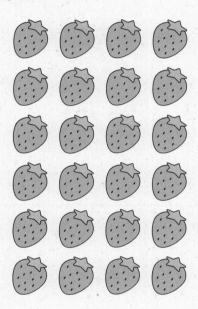

$$3 \times (2 \times 4)$$

$$(3 \times 2) \times 4$$

2. 4 groups of 3 × 2

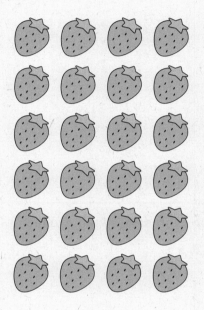

$$(4 \times 3) \times 2$$

$$4 \times (3 \times 2)$$

3. 2 groups of 3 × 5

4. 3 groups of 2 × 5

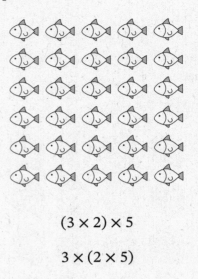

$(2 \times 3) \times 5$

$2 \times (3 \times 5)$

$(3 \times 2) \times 5$

$3 \times (2 \times 5)$

Rewrite the expressions with two factors. Then find the products.

5. $(4 \times 2) \times 5$

6. $4 \times (2 \times 5)$

7. $(3 \times 2) \times 2$

8. $3 \times (2 \times 2)$

9. $(2 \times 2) \times 5$

10. $2 \times (2 \times 5)$

11. $(5 \times 1) \times 3$

12. $5 \times (1 \times 3)$

Place parentheses in the equations to simplify and complete the problem. The first one has been done for you.

13. $14 \times 2 = (7 \times 2) \times 2$
$= 7 \times (2 \times 2)$
$= 7 \times \underline{\quad 4 \quad}$
$= \underline{\quad 28 \quad}$

14. $3 \times 16 = 3 \times (2 \times 8)$
$= 3 \times 2 \times 8$
$= \underline{\quad\quad} \times 8$
$= \underline{\quad\quad}$

15. $2 \times 15 = 2 \times (3 \times 5)$
$= 2 \times 3 \times 5$
$= \underline{\quad\quad} \times \underline{\quad\quad}$
$= \underline{\quad\quad}$

16. $14 \times 3 = 7 \times 2 \times 3$
$= 7 \times 2 \times 3$
$= \underline{\quad\quad} \times \underline{\quad\quad}$
$= \underline{\quad\quad}$

17. $15 \times 3 = 5 \times 3 \times 3$
$= 5 \times 3 \times 3$
$= \underline{\quad\quad} \times \underline{\quad\quad}$
$= \underline{\quad\quad}$

18. $2 \times 16 = 4 \times 4 \times 2$
$= 4 \times 4 \times 2$
$= \underline{\quad\quad} \times \underline{\quad\quad}$
$= \underline{\quad\quad}$

19. Gabe finds the answer to 16 × 4 by thinking about 8 × 8. Explain his strategy.

Name _____

Fill in the blanks to make the equations true.

1. $3 \times (2 \times 7) = (\underline{\hspace{1cm}} \times 2) \times 7$

2. $(4 \times 2) \times 5 = 4 \times (\underline{\hspace{1cm}} \times 5)$

Place parentheses in the equations to simplify and complete the problems.

3. $14 \times 4 = 7 \times 2 \times 4$

   $= 7 \times 2 \times 4$

   $= \underline{\hspace{1cm}} \times \underline{\hspace{1cm}}$

   $= \underline{\hspace{1cm}}$

4. $5 \times 12 = 5 \times 2 \times 6$

   $= 5 \times 2 \times 6$

   $= \underline{\hspace{1cm}} \times \underline{\hspace{1cm}}$

   $= \underline{\hspace{1cm}}$

Name _____

Complete each equation.

1. $(15 - 5) + 7 =$ _____

 $15 - (5 + 7) =$ _____

2. $(14 \div 2) + 5 =$ _____

 $14 \div (2 + 5) =$ _____

3. $(13 + 7) - 3 =$ _____

 $13 + (7 - 3) =$ _____

4. $6 \times (7 + 1) =$ _____

 $(6 \times 7) + 1 =$ _____

5. _____ $= (18 \div 6) + 3$

 _____ $= 18 \div (6 + 3)$

6. $15 - (5 \times 2) =$ _____

 $(15 - 5) \times 2 =$ _____

Complete each equation. Circle the pairs that have the same value for both equations.

7. $8 + (6 + 4) =$ _____

 $(8 + 6) + 4 =$ _____

8. $(3 \times 4) \times 2 =$ _____

 $3 \times (4 \times 2) =$ _____

9. $(5 \times 2) \times 4 =$ _____

 $5 \times (2 \times 4) =$ _____

10. $6 \times (5 - 4) =$ _____

 $(6 \times 5) - 4 =$ _____

11. $10 - (5 + 2) =$ _____

 $(10 - 5) + 2 =$ _____

12. $4 \times (2 \times 2) =$ _____

 $(4 \times 2) \times 2 =$ _____

13. $(6 + 3) \times 2 =$ _____

 $6 + (3 \times 2) =$ _____

14. $(12 \div 4) \times 3 =$ _____

 $12 \div (4 \times 3) =$ _____

15. Create two different expressions by grouping different parts of $3 \times 4 + 5$ by using parentheses. Then find their values.

$3 \times 4 + 5 = $ _____

$3 \times 4 + 5 = $ _____

Use parentheses to make each equation true.

16. $16 - 8 + 7 = 15$

17. $16 - 8 + 7 = 1$

18. $7 = 7 \times 4 - 3$

19. $25 = 7 \times 4 - 3$

20. $50 \div 10 \times 5 = 25$

21. $50 \div 10 \times 5 = 1$

22. $5 \times 6 \div 3 = 10$

23. $6 = 3 \times 8 \div 4$

24. $10 = 20 \div 10 \times 5$

25. $56 = 7 + 7 \times 7$

---

26. Amy and Eva both find the value of $20 \div (2 + 3)$.

 • Amy says the value is 4.

 • Eva says the value is 13.

     a. Who is correct? Explain how you know.

     b. What mistake did the other student make?

---

27. Ray says the value of $3 \times 8 \div 4$ is 6 no matter where he puts the parentheses. Is he correct? Place parentheses around different numbers to explain his thinking.

Name

Use parentheses to make the equation true.

1. $2 + 8 \times 7 = 70$

2. $2 + 8 \times 7 = 58$

3. Zara and Luke both find the value of $28 + (14 \div 7)$.

   • Zara says the value is 30.

   • Luke says the value is 6.

   a. Who is correct? Explain how you know.

   b. What mistake did the other student make?

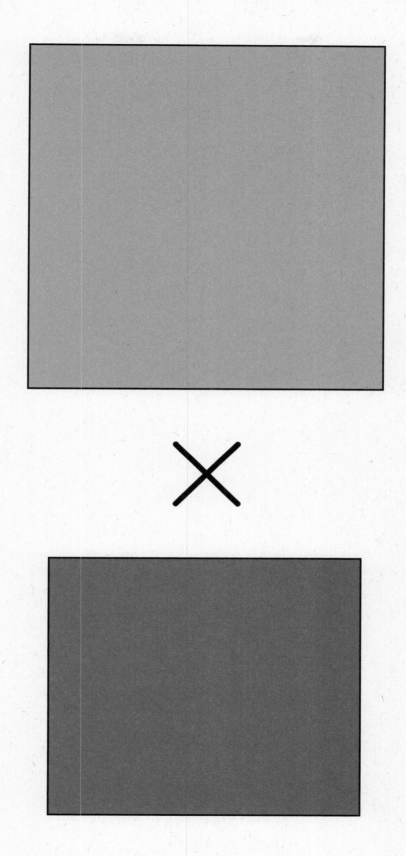

Name

1. Complete the equations in the number bond to find $56 \div 7$. Use each part of the array to help divide.

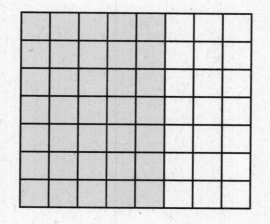

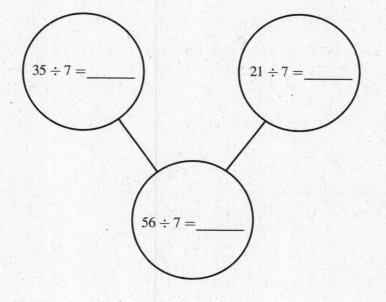

$$35 \div 7 = \underline{\hspace{1cm}}$$

$$21 \div 7 = \underline{\hspace{1cm}}$$

$$56 \div 7 = \underline{\hspace{1cm}}$$

Divide by using the break apart and distribute strategy.

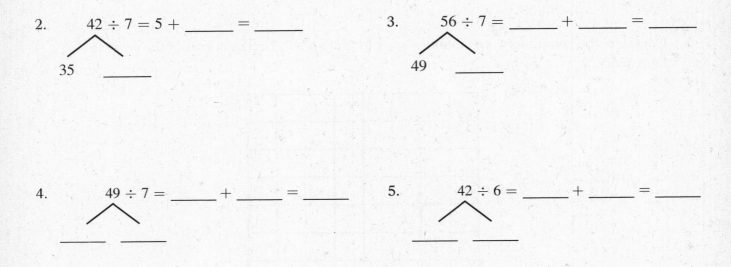

2. $42 \div 7 = 5 +$ _____ = _____

35   _____

3. $56 \div 7 =$ _____ + _____ = _____

49   _____

4. $49 \div 7 =$ _____ + _____ = _____

_____ _____

5. $42 \div 6 =$ _____ + _____ = _____

_____ _____

6. Find $56 \div 8$ by using the break apart and distribute strategy in two different ways.

$56 \div 8 =$ _____

_____ _____

$56 \div 8 =$ _____

_____ _____

Divide by using the break apart and distribute strategy.

7. $72 \div 6 =$ _____

_____ _____

8. $96 \div 8 =$ _____

_____ _____

Use the Read–Draw–Write process to solve the problem.

9. Robin has 96 cat stickers. She puts them in rows of 8. How many rows of cat stickers does Robin have?

_____

Name

Divide by using the break apart and distribute strategy.

1. $78 \div 6 =$ _____ $+$ _____ $=$ _____

2. $72 \div 8 =$ _____ $+$ _____ $=$ _____

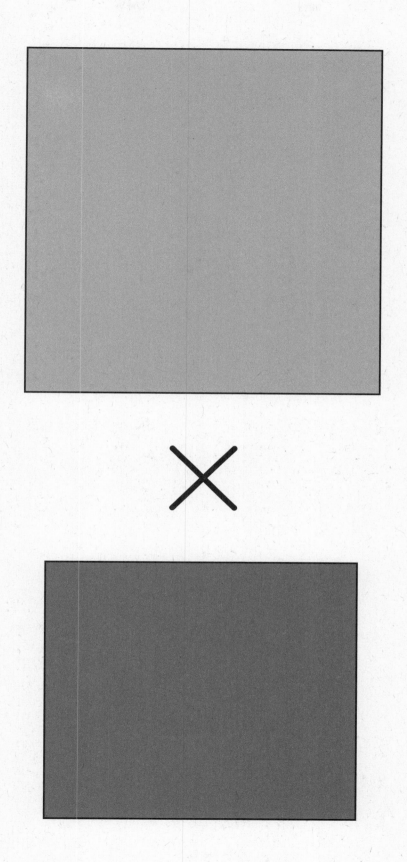

Name

Use the Read–Draw–Write process to solve the problem. Use a letter to represent the unknown.

1.   The juice boxes in the cooler are in 4 rows and 7 columns. How many juice boxes are there?

Use the Read–Draw–Write process to solve the problem. Use a letter to represent the unknown.

2.   49 milk cartons are arranged on a tray in 7 columns. How many rows of milk cartons are there?

Name

Use the Read–Draw–Write process to solve each problem. Use a letter to represent the unknown in each.

1. 7 teams compete in a relay race. Each team has 5 runners. How many runners compete in all?

2. Shen puts his rubber bands into 6 equal piles. He has a total of 48 rubber bands. How many rubber bands are in each pile?

3. Zara sets up chairs for a concert. She puts them in 7 rows. Each row has 9 chairs. How many chairs does she set up altogether?

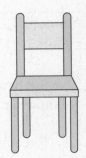

4. Mrs. Smith sets up toy trains. The train cars are arranged in 6 rows and 7 columns. How many train cars are there?

5. At the store, a display of bricks has 7 equal columns. There are 56 bricks in all. How many bricks are in each column?

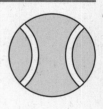

6. Coach Endo wants to put his tennis balls equally into cans. There are 64 tennis balls. Each can holds 8 tennis balls. How many cans does he need?

Name _____

Use the Read–Draw–Write process to solve the problem. Use a letter to represent the unknown.

Mr. Davis arranges 72 chairs into 8 equal rows. How many chairs are in each row?

Name _____

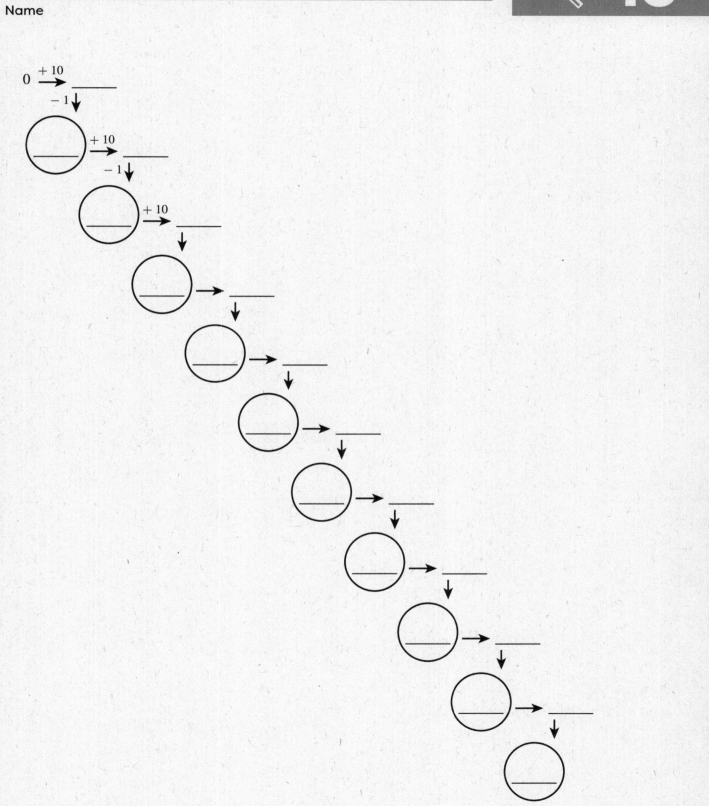

Name

1. Fill in the blanks to help you skip-count by nines.

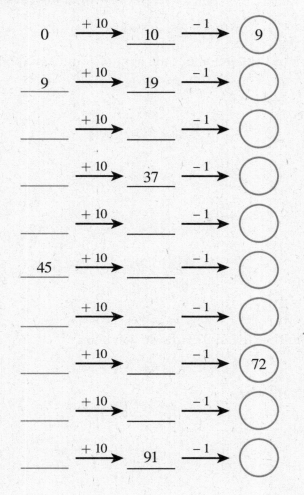

2. How does adding 10 and subtracting 1 help you skip-count by 9?

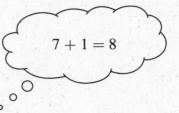

$$7 + 1 = 8$$

3. Oka writes $8 \times 9 = 71$.

   a. She checks her work by thinking about the sum of 7 and 1. Explain Oka's strategy.

   b. Did Oka correctly multiply 8 and 9? How do you know?

Find each product. Describe the strategy that you used.

4. 7×9

_____

5. 6×9

_____

6. 9×9

_____

7. How can you tell if your answers to problems 4–6 are correct?

Name _____

Fill in the blanks to find the next nines fact.

1. $8 \times 9 = 72$

    What is 10 more than 72? _____

    What is 1 less than that? _____

    $9 \times 9 =$ _____

2. $6 \times 9 = 54$

    What is 10 more than 54? _____

    What is 1 less than that? _____

    $7 \times 9 =$ _____

3. Describe the pattern used in problems 1 and 2.

Match each expression to the word problem that it represents.

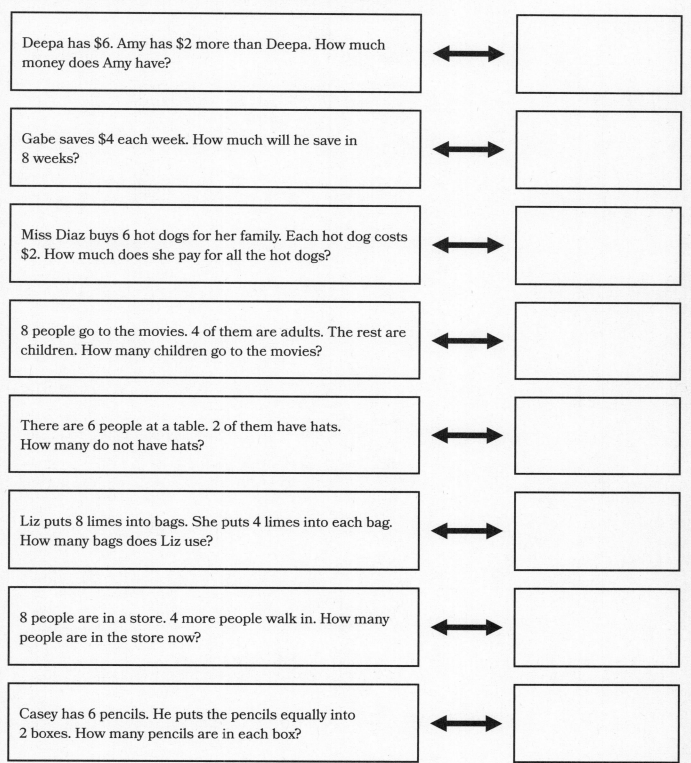

Deepa has $6. Amy has $2 more than Deepa. How much money does Amy have?	↔
Gabe saves $4 each week. How much will he save in 8 weeks?	↔
Miss Diaz buys 6 hot dogs for her family. Each hot dog costs $2. How much does she pay for all the hot dogs?	↔
8 people go to the movies. 4 of them are adults. The rest are children. How many children go to the movies?	↔
There are 6 people at a table. 2 of them have hats. How many do not have hats?	↔
Liz puts 8 limes into bags. She puts 4 limes into each bag. How many bags does Liz use?	↔
8 people are in a store. 4 more people walk in. How many people are in the store now?	↔
Casey has 6 pencils. He puts the pencils equally into 2 boxes. How many pencils are in each box?	↔

Name

Name _____

1. Draw lines to match. Fill in the blanks to complete the expressions.

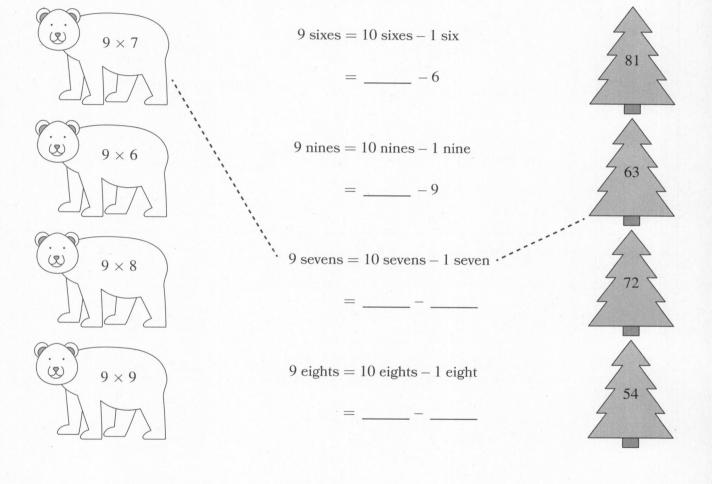

9 sixes = 10 sixes − 1 six

= _____ − 6

9 nines = 10 nines − 1 nine

= _____ − 9

9 sevens = 10 sevens − 1 seven

= _____ − _____

9 eights = 10 eights − 1 eight

= _____ − _____

Fill in the blanks to find the total value of the shaded parts.

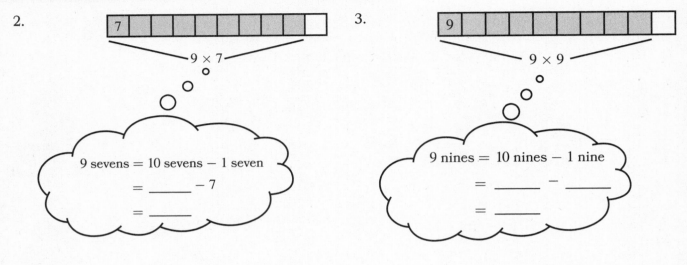

2.

9×7

9 sevens = 10 sevens − 1 seven

= _____ − 7

= _____

3.

9×9

9 nines = 10 nines − 1 nine

= _____ − _____

= _____

---

4. Use the table for parts (a)–(d).

a. Multiply. Then add the tens digit and the ones digit of each product.

$1 \times 9 = \underline{\ \ 9\ \ }$	$\underline{\ \ 0\ \ } + \underline{\ \ 9\ \ } = \underline{\ \ 9\ \ }$
$2 \times 9 = \underline{\ \ 18\ \ }$	$\underline{\ \ 1\ \ } + \underline{\ \ 8\ \ } = \underline{\ \ \ \ }$
$3 \times 9 = \underline{\ \ \ \ }$	$\underline{\ \ \ \ } + \underline{\ \ \ \ } = \underline{\ \ \ \ }$
$4 \times 9 = \underline{\ \ \ \ }$	$\underline{\ \ \ \ } + \underline{\ \ \ \ } = \underline{\ \ \ \ }$
$5 \times 9 = \underline{\ \ \ \ }$	$\underline{\ \ \ \ } + \underline{\ \ \ \ } = \underline{\ \ \ \ }$
$6 \times 9 = \underline{\ \ \ \ }$	$\underline{\ \ \ \ } + \underline{\ \ \ \ } = \underline{\ \ \ \ }$
$7 \times 9 = \underline{\ \ \ \ }$	$\underline{\ \ \ \ } + \underline{\ \ \ \ } = \underline{\ \ \ \ }$
$8 \times 9 = \underline{\ \ \ \ }$	$\underline{\ \ \ \ } + \underline{\ \ \ \ } = \underline{\ \ \ \ }$
$9 \times 9 = \underline{\ \ \ \ }$	$\underline{\ \ \ \ } + \underline{\ \ \ \ } = \underline{\ \ \ \ }$
$10 \times 9 = \underline{\ \ \ \ }$	$\underline{\ \ \ \ } + \underline{\ \ \ \ } = \underline{\ \ \ \ }$

b. Look at the tens place in each product. What pattern do you notice?

c. Look at the ones place in each product. What pattern do you notice?

d. What is the sum of the digits in each product? How can the sum of the digits help you check your work with nines facts?

---

5. James buys a box of chalk. It has 9 rows with 4 sticks of chalk in each row. He uses 10 fours to find the total number of chalk sticks.

a. Draw a model to represent James's strategy.

b. Explain James's strategy and find the total number of chalk sticks.

_____

_____

_____

_____

6. Carla finds 9×8 by subtracting 1 eight from 10 eights. Deepa finds 9×8 by subtracting 1 nine from 10 nines. Whose strategy is correct? Explain how you know.

Carla

$9 \times 8 = 10$ eights $- 1$ eight
    $= 80 - 8$
    $= 72$

Deepa

$9 \times 8 = 10$ nines $- 1$ nine
    $= 90 - 9$
    $= 81$

_____

_____

_____

_____

7. Use the expressions for parts (a) and (b).

    a. Circle the expressions that are equal to 9×7.

$(5 \times 7) + (4 \times 7)$	10 nines $- 1$ nine
$54 + 10 - 1$	$(10 \times 7) - (1 \times 7)$

    b. Choose one expression you circled. Explain how you know it is equal to 9×7.

_____

_____

_____

_____

Name

1. David writes $9 \times 6 = 54$. Show two strategies you could use to check his answer.

2. Circle the expressions that are equal to 9×6.

$(9 \times 5) + 9$ $(8 \times 6) + 8$

$(10 \times 6) - (1 \times 6)$ $(5 \times 6) + (4 \times 6)$

Match each expression to the word problem that it represents.

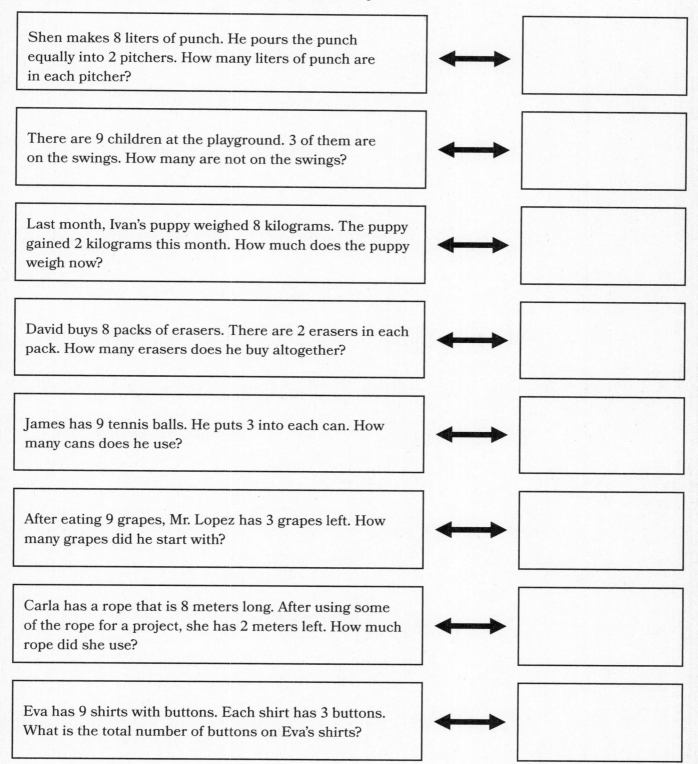

Shen makes 8 liters of punch. He pours the punch equally into 2 pitchers. How many liters of punch are in each pitcher?

There are 9 children at the playground. 3 of them are on the swings. How many are not on the swings?

Last month, Ivan's puppy weighed 8 kilograms. The puppy gained 2 kilograms this month. How much does the puppy weigh now?

David buys 8 packs of erasers. There are 2 erasers in each pack. How many erasers does he buy altogether?

James has 9 tennis balls. He puts 3 into each can. How many cans does he use?

After eating 9 grapes, Mr. Lopez has 3 grapes left. How many grapes did he start with?

Carla has a rope that is 8 meters long. After using some of the rope for a project, she has 2 meters left. How much rope did she use?

Eva has 9 shirts with buttons. Each shirt has 3 buttons. What is the total number of buttons on Eva's shirts?

Name _____

Fill in each box with your assigned number. Then draw a picture to represent each problem and complete each statement and equation.

1. ☐ groups of 1 is _____ .

   ☐ × _____ = _____

2. 1 group of ☐ is _____ .

   _____ × ☐ = _____

3. ☐ divided into groups of 1 is _____ groups.

   ☐ ÷ _____ = _____

4. ☐ divided into 1 group is _____ .

   ☐ ÷ _____ = _____

5. ☐ divided into ☐ equal groups is _____ in each group.

   ☐ ÷ ☐ = _____

Complete the statements and equations.

6. 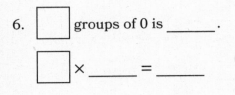 groups of 0 is _____ .

7. 0 groups of 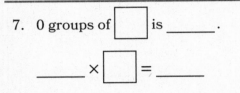 is _____ .

8. Complete the statement and equation.

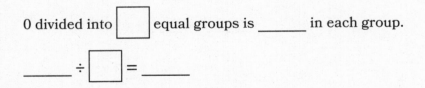

15

Name

Use the equal groups pictures to fill in the blanks.

1. 4 groups of 1 is _____ .

   $4 \times 1 =$ _____

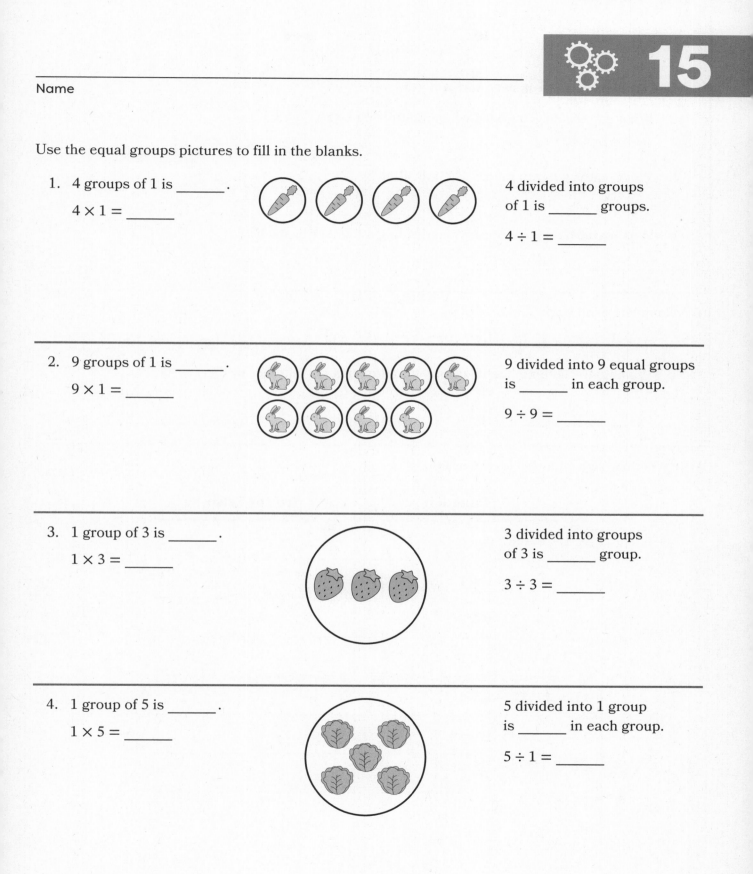

   4 divided into groups of 1 is _____ groups.

   $4 \div 1 =$ _____

2. 9 groups of 1 is _____ .

   $9 \times 1 =$ _____

   9 divided into 9 equal groups is _____ in each group.

   $9 \div 9 =$ _____

3. 1 group of 3 is _____ .

   $1 \times 3 =$ _____

   3 divided into groups of 3 is _____ group.

   $3 \div 3 =$ _____

4. 1 group of 5 is _____ .

   $1 \times 5 =$ _____

   5 divided into 1 group is _____ in each group.

   $5 \div 1 =$ _____

5. Use problems 1–4 to answer parts (a)–(c).

   a. What pattern do you notice when multiplying by 1?

   b. What pattern do you notice when dividing by 1?

   c. What pattern do you notice when dividing a number by itself?

---

6. Complete each statement.

   a. 0 groups of any number is _____ .   b. Any number of groups of 0 is _____ .

   c. What do the statements tell us about multiplying any number by 0?

Write whether each equation is true or false.

Equation	True or False
7. $8 \times 0 = 0$	
8. $7 \div 1 = 1$	
9. $1 \times 6 = 6$	
10. $5 \div 5 = 1$	
11. $9 \times 1 = 1$	
12. $0 \times 10 = 0$	
13. $3 \div 0 = 0$	

14. Choose one of the false equations from problems 7–13 and explain why it is false.

_____

_____

_____

_____

15. Complete each statement.

    a.   0 divided by any number except 0 is _____ .

    b.   We cannot divide by _____ .

Name

Fill in the blanks to make the equations true.

1. _____ $\times 1 = 5$

2. $6 \times$ _____ $= 6$

3. _____ $\div 7 = 0$

4. $5 \times$ _____ $= 0$

5. $1 = 9 \div$ _____

Circle true or false for each statement.

6. Any number except 0 divided by itself is 1.   True   False

7. 0 divided by any number except 0 is 0.   True   False

8. Any number divided by 1 is 1.   True   False

Sprint

Complete the equations.

1.	$2 \times 3 =$ _____	
2.	$6 \div 3 =$ _____	

A

Number Correct: _____

Complete the equations.

1.	$1 \times 3 =$ _____	
2.	$2 \times 3 =$ _____	
3.	$3 \times 3 =$ _____	
4.	$3 \div 3 =$ _____	
5.	$6 \div 3 =$ _____	
6.	$9 \div 3 =$ _____	
7.	$4 \times 3 =$ _____	
8.	$5 \times 3 =$ _____	
9.	$6 \times 3 =$ _____	
10.	$12 \div 3 =$ _____	
11.	$15 \div 3 =$ _____	
12.	$18 \div 3 =$ _____	
13.	$7 \times 3 =$ _____	
14.	$8 \times 3 =$ _____	
15.	$9 \times 3 =$ _____	
16.	$21 \div 3 =$ _____	
17.	$24 \div 3 =$ _____	
18.	$27 \div 3 =$ _____	
19.	$10 \times 3 =$ _____	
20.	$1 \times 3 =$ _____	
21.	$30 \div 3 =$ _____	
22.	$3 \div 3 =$ _____	

23.	$3 \times$ _____ $= 9$	
24.	$3 \times$ _____ $= 15$	
25.	$9 \div 3 =$ _____	
26.	$15 \div 3 =$ _____	
27.	$3 \times$ _____ $= 21$	
28.	$3 \times$ _____ $= 27$	
29.	$21 \div 3 =$ _____	
30.	$27 \div 3 =$ _____	
31.	$3 \times$ _____ $= 3$	
32.	$3 \times$ _____ $= 0$	
33.	$3 \div 3 =$ _____	
34.	$9 \div 3 =$ _____	
35.	$3 \times$ _____ $= 6$	
36.	_____ $\times 3 = 12$	
37.	$3 \times$ _____ $= 18$	
38.	_____ $\times 3 = 24$	
39.	$3 \times$ _____ $= 30$	
40.	_____ $\div 3 = 2$	
41.	$12 \div$ _____ $= 3$	
42.	_____ $\div 3 = 6$	
43.	$24 \div$ _____ $= 3$	
44.	_____ $\div 3 = 10$	

B

Number Correct: _____

Improvement: _____

Complete the equations.

1.	$1 \times 3 =$ _____	
2.	$2 \times 3 =$ _____	
3.	$3 \times 3 =$ _____	
4.	$3 \div 3 =$ _____	
5.	$6 \div 3 =$ _____	
6.	$9 \div 3 =$ _____	
7.	$3 \times 3 =$ _____	
8.	$4 \times 3 =$ _____	
9.	$5 \times 3 =$ _____	
10.	$9 \div 3 =$ _____	
11.	$12 \div 3 =$ _____	
12.	$15 \div 3 =$ _____	
13.	$6 \times 3 =$ _____	
14.	$7 \times 3 =$ _____	
15.	$8 \times 3 =$ _____	
16.	$18 \div 3 =$ _____	
17.	$21 \div 3 =$ _____	
18.	$24 \div 3 =$ _____	
19.	$9 \times 3 =$ _____	
20.	$10 \times 3 =$ _____	
21.	$27 \div 3 =$ _____	
22.	$30 \div 3 =$ _____	

23.	$3 \times$ _____ $= 6$	
24.	$3 \times$ _____ $= 12$	
25.	$6 \div 3 =$ _____	
26.	$12 \div 3 =$ _____	
27.	$3 \times$ _____ $= 18$	
28.	$3 \times$ _____ $= 24$	
29.	$18 \div 3 =$ _____	
30.	$24 \div 3 =$ _____	
31.	$3 \times$ _____ $= 3$	
32.	$3 \times$ _____ $= 0$	
33.	$3 \div 3 =$ _____	
34.	$6 \div 3 =$ _____	
35.	$3 \times$ _____ $= 9$	
36.	_____ $\times 3 = 15$	
37.	$3 \times$ _____ $= 21$	
38.	_____ $\times 3 = 27$	
39.	$3 \times$ _____ $= 30$	
40.	_____ $\div 3 = 3$	
41.	$15 \div$ _____ $= 3$	
42.	_____ $\div 3 = 7$	
43.	$27 \div$ _____ $= 3$	
44.	_____ $\div 3 = 10$	

16

Name _____

1. Read about Mr. Davis's class and then answer the question about Miss Diaz's class.

Mr. Davis has 21 students. Mr. Davis put his students into pairs. He asked his class whether there is an even or odd number of students and to explain how they know.

Ray said, "There is an odd number. No matter how you rearrange the groups, one student does not have a partner."

James said, "There is an odd number. When I count by twos, I do not say 21."

Jayla said, "There is an odd number. I cannot write a doubles fact. I have to think of doubles plus one: $10 + 10 + 1 = 21$."

Miss Diaz has 18 students. Miss Diaz put her students into pairs. Does Miss Diaz have an even number of students or an odd number of students? Draw a picture and explain your reasoning.

2. Write the products in the squares to complete the table.

×	1	2	3	4	5	6	7	8	9	10
1										
2										
3										
4										
5										
6										
7										
8										
9										
10										

3. When the product is even, can either factor be odd? How do you know?

_____

_____

_____

4. When the product is odd, can either factor be even? How do you know?

_____

_____

_____

5. Break apart the larger fact into smaller facts to find 15×4.

16

Name

Use your completed multiplication table to answer problems 1–5.

1. Decide whether each pattern is true or false. Write an equation that supports your decision.

Pattern	True or False	Equation
Even times even equals odd.		
Odd times odd equals even.		
Even times odd equals odd.		

2. Circle the product 28 in the multiplication table. Explain why the product 28 is in the table more than once.

3. How does the table help you see that any number multiplied by 1 equals itself?

4. Eva says, "I can use the multiplication table to find 7 × 14."

   David says, "But 7 × 14 is not in the table."

   Explain how Eva might use the table to find 7 × 14.

5. How can you use the table to find $48 \div 6$?

Name

1. Use the multiplication table to find 8 × 12.

×	1	2	3	4	5	6	7	8	9	10
1	1	2	3	4	5	6	7	8	9	10
2	2	4	6	8	10	12	14	16	18	20
3	3	6	9	12	15	18	21	24	27	30
4	4	8	12	16	20	24	28	32	36	40
5	5	10	15	20	25	30	35	40	45	50
6	6	12	18	24	30	36	42	48	54	60
7	7	14	21	28	35	42	49	56	63	70
8	8	16	24	32	40	48	56	64	72	80
9	9	18	27	36	45	54	63	72	81	90
10	10	20	30	40	50	60	70	80	90	100

2. Carla says 3 × 233 = 626. Use what you learned about odd times odd to explain why Carla is wrong.

Name _____

1. Liz puts fruit in bags to sell at the market.

   a. Liz puts 9 apples in each bag. Complete the table to show the total number of apples in bags.

Number of Bags	1	2	3	4	5
Total Number of Apples	9	18			

   b. Liz puts 5 pears in each bag. Complete the table to show the number of bags of pears.

Total Number of Pears	5	10	15	20	25
Number of Bags	1			4	

2. Liz fills bags with plums. She puts the same number of plums in each bag. Complete the table to show the total number of plums in bags.

Number of Bags	1	2	3	4	5	6
Total Number of Plums	4	8				

3. Look for a pattern to complete the table.

Input	42	36	30	24	18	12
Output	7			4		2

   Pattern: _____

4.

Input	Output
2	16
3	24
5	
7	
8	64
10	

Pattern: _____

5. Use the table to help you find 12 × 7.

1	7
2	14
4	28
8	56

Name _____

1. A triangle has 3 sides. Complete the table.

Number of Triangles	1	2	3	4	5	6	7
Total Number of Sides	3	6					

2. Liz sells bags of peaches. Each bag has the same number of peaches.

a. Complete the table.

Number of Bags	1	2	3	4	5	6
Total Number of Peaches	6	12			30	

b. How many peaches are there in 8 bags? How do you know?

3. Mr. Endo puts carrots on lunch trays.

a. Complete the table.

Number of Carrots	5	10	15	20	25	30	35
Total Number of Lunch Trays	1		3				7

b. How many lunch trays does Mr. Endo need if he has 45 carrots?

Use the pattern to complete each table.

4. Pattern: Multiply the input by 4

Input	2	3	4	5	6	7	8
Output	8						

5. Pattern: Multiply the input by 9

Input	Output
3	
4	
6	
8	
9	

6. Pattern: Divide the input by 8

Input	Output
72	9
56	
40	
32	
16	

Write the pattern and complete each table.

7. Pattern: _____

Input	2	4	5	7	9	10
Output	14		35		63	

8. Pattern: _____

Input	Output
18	9
14	
12	6
8	
6	3
4	

9. Pattern: _____

Input	Output
30	5
6	
42	7
12	2
54	
18	

Name _____

Write the pattern and complete the table.

Pattern: _____

Input	Output
1	7
2	14
3	21
	42
5	
	28

7×8	$56 \div 7$	$56 \div 8$

There are 56 flowers.	There are 7 piles of hard hats.	There are 56 stuffed bears.

There are 8 hard hats in each pile.	The bears are divided evenly among 7 shelves.	The flowers are put into pots with 8 flowers in each pot.

How many bears are on each shelf?	How many pots are there?	How many hard hats are there in all?

Name _____

1. Use the picture to write a word problem that can be represented with the expression 4 × 5.

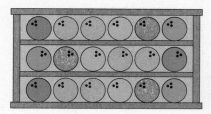

_____

_____

_____

_____

2. Use the picture to write a word problem that can be represented with the expression 18 ÷ 6.

_____

_____

_____

_____

18

Name

1. Write a word problem that can be represented with the expression 5 × 9. Use the picture to help you.

_____

_____

_____

_____

2. Write a word problem that can be represented with the expression 4×9. Use the picture to help you.

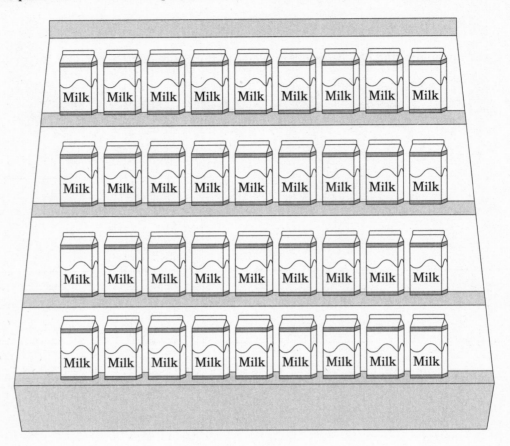

_____

_____

_____

_____

3. Write a word problem that can be represented with the expression 27 ÷ 9. Use the picture to help you.

_____

_____

_____

_____

4. Write a word problem that can be represented with the expression 18 ÷ 2. Use the picture to help you.

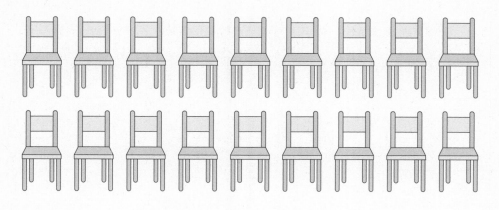

_____

_____

_____

_____

18

Name _____

Write a word problem that can be represented with the expression 18 ÷ 3.
Use the picture to help you.

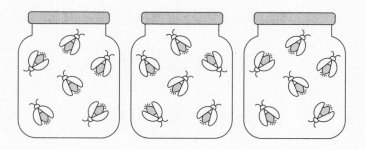

_____

_____

_____

_____

Name _____

1. There are 9 packs of markers in the supply box.

   Each pack has 8 markers.

   After students take some of the markers, there are 19 markers left.

   How many markers did students take?

   a. Draw to represent the problem. Use a letter to represent each unknown.

   b. Estimate the number of markers students took.

   c. Solve the problem with equations and write a solution statement.

   d. How do you know your solution is reasonable? Explain.

2. David had $173 in the bank.

   He earned $9 each week and put that money in the bank too.

   Now David has $218 in the bank.

   For how many weeks did David earn money?

   a. Draw to represent the problem. Use a letter to represent each unknown.

   b. Estimate how many weeks David earned money. Use the questions to help you.

   About how much money did David start with?

   About how much money does David have now?

   About how much total money did David earn?

   So about how many weeks did David earn money?

   c. Solve the problem with equations and write a solution statement.

   d. How do you know your solution is reasonable? Explain.

Name

Use the Read–Draw–Write process to solve each problem.

1. Deepa buys 4 packs of plates for a party. Each pack has 9 plates.

Deepa and her friends use 27 plates during the party.

How many plates are left?

a. Draw to represent the problem. Use a letter to represent each unknown.

b. Estimate how many plates are left. Use the questions to help you.

About how many plates does Deepa buy?

About how many plates do Deepa and her friends use?

So about how many plates are left?

c. Solve the problem. Write equations and a solution statement.

d. How do you know your answer is reasonable? Use your estimate from part (b) to help you explain.

2. 120 students participate in Field Day activities.

   48 students play beanbag toss, and the rest of the students play kickball.

   Each kickball team has 9 players. How many kickball teams are there?

   a. Draw to represent the problem. Use a letter to represent each unknown.

   b. Estimate how many teams there are. Use the questions to help you.

      About how many students play kickball?

      About how many players are on each team?

      So about how many teams are there?

   c. Solve the problem. Write equations and a solution statement.

   d. How do you know your answer is reasonable? Use your estimate from part (b) to help you explain.

3. Mr. Lopez buys 9 packs of 4 markers. He puts an equal number of markers on each table.

   There are 6 tables.

   How many markers does Mr. Lopez put on each table?

   a. Draw to represent the problem. Use a letter to represent each unknown.

   b. Estimate.

   c. Solve the problem. Write equations and a solution statement.

   d. How do you know your answer is reasonable? Use your estimate from part (b) to help you explain.

4. Ray has $48 in his wallet and $33 in his piggy bank. He uses all the money to buy books. Each book costs $9. How many books does Ray buy?

    a. Draw to represent the problem. Use a letter to represent each unknown.

    b. Estimate.

    c. Solve the problem. Write equations and a solution statement.

    d. How do you know your answer is reasonable? Use your estimate from part (b) to help you explain.

Name

Amy's book is 96 pages long.

She has already read 42 pages.

How many nights will it take Amy to finish the book if she reads 9 pages each night?

    a. Draw to represent the problem. Use a letter to represent each unknown.

    b. Estimate how many nights it will take Amy to finish the book.

    c. Solve the problem. Write equations and a solution statement.

    d. How do you know your solution is reasonable? Explain, using your estimate from part (b).

Name _____

Complete the equations. Use the place value disks to help you.

1.

2 × 4 ones = _____ ones

2 × 4 = _____

2.

2 × 4 tens = _____ tens

2 × 40 = _____

Complete the equations. Use the place value charts to help you.

3.
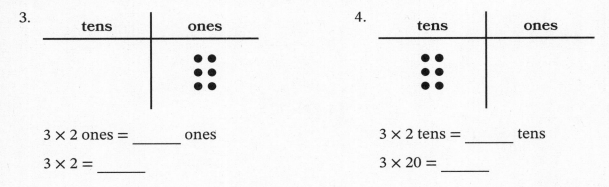

tens	ones

3 × 2 ones = _____ ones

3 × 2 = _____

4.

tens	ones

3 × 2 tens = _____ tens

3 × 20 = _____

5.
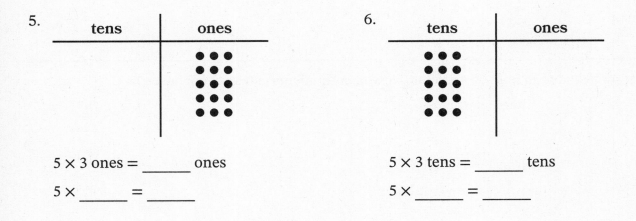

tens	ones

5 × 3 ones = _____ ones

5 × _____ = _____

6.

tens	ones

5 × 3 tens = _____ tens

5 × _____ = _____

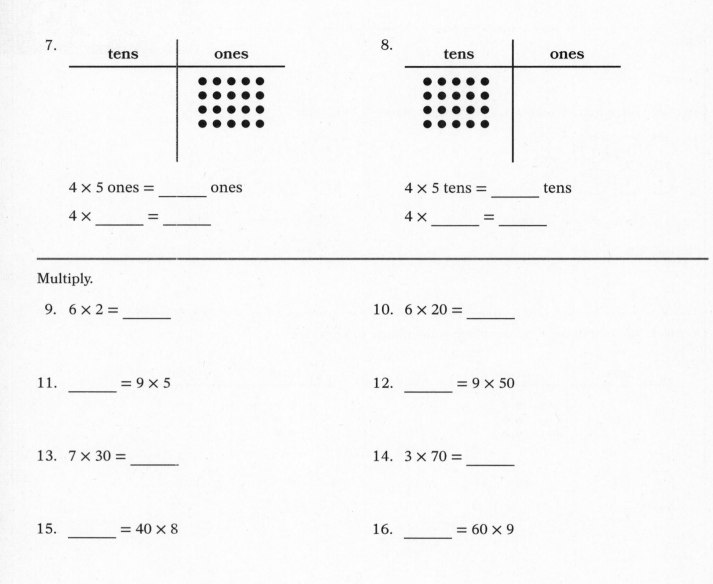

7.

tens	ones

4 × 5 ones = _____ ones

4 × _____ = _____

8.

tens	ones

4 × 5 tens = _____ tens

4 × _____ = _____

Multiply.

9. 6 × 2 = _____

10. 6 × 20 = _____

11. _____ = 9 × 5

12. _____ = 9 × 50

13. 7 × 30 = _____

14. 3 × 70 = _____

15. _____ = 40 × 8

16. _____ = 60 × 9

17. Each school bus can hold 60 students. How many students can 4 school buses hold?

Name _____

Complete the equations. Use the place value charts to help you.

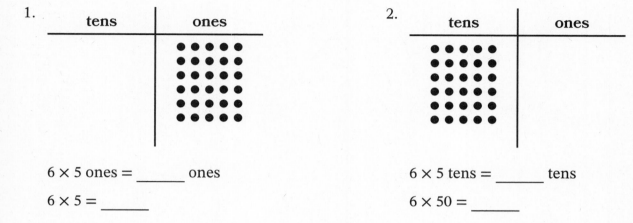

1.

tens	ones

6×5 ones = _____ ones

$6 \times 5 =$ _____

2.

tens	ones

6×5 tens = _____ tens

$6 \times 50 =$ _____

3. Each page in Miss Diaz's sticker book has 40 stickers. How many stickers are there on 9 pages?

Sprint

Complete the equations.

1.	$2 \times 4 = \underline{\hspace{2cm}}$	
2.	$8 \div 4 = \underline{\hspace{2cm}}$	

A

Complete the equations.

1.	$1 \times 4 =$ _____	
2.	$2 \times 4 =$ _____	
3.	$3 \times 4 =$ _____	
4.	$4 \div 4 =$ _____	
5.	$8 \div 4 =$ _____	
6.	$12 \div 4 =$ _____	
7.	$4 \times 4 =$ _____	
8.	$5 \times 4 =$ _____	
9.	$6 \times 4 =$ _____	
10.	$16 \div 4 =$ _____	
11.	$20 \div 4 =$ _____	
12.	$24 \div 4 =$ _____	
13.	$7 \times 4 =$ _____	
14.	$8 \times 4 =$ _____	
15.	$9 \times 4 =$ _____	
16.	$28 \div 4 =$ _____	
17.	$32 \div 4 =$ _____	
18.	$36 \div 4 =$ _____	
19.	$10 \times 4 =$ _____	
20.	$1 \times 4 =$ _____	
21.	$40 \div 4 =$ _____	
22.	$4 \div 4 =$ _____	

23.	$4 \times$ _____ $= 12$	
24.	$4 \times$ _____ $= 20$	
25.	$12 \div 4 =$ _____	
26.	$20 \div 4 =$ _____	
27.	$4 \times$ _____ $= 28$	
28.	$4 \times$ _____ $= 36$	
29.	$28 \div 4 =$ _____	
30.	$36 \div 4 =$ _____	
31.	$4 \times$ _____ $= 4$	
32.	$4 \times$ _____ $= 0$	
33.	$4 \div 4 =$ _____	
34.	$12 \div 4 =$ _____	
35.	$4 \times$ _____ $= 8$	
36.	_____ $\times 4 = 16$	
37.	$4 \times$ _____ $= 24$	
38.	_____ $\times 4 = 32$	
39.	$4 \times$ _____ $= 40$	
40.	_____ $\div 4 = 2$	
41.	$16 \div$ _____ $= 4$	
42.	_____ $\div 4 = 6$	
43.	$32 \div$ _____ $= 4$	
44.	_____ $\div 4 = 10$	

B

Complete the equations.

1.	$1 \times 4 =$ _____	
2.	$2 \times 4 =$ _____	
3.	$3 \times 4 =$ _____	
4.	$4 \div 4 =$ _____	
5.	$8 \div 4 =$ _____	
6.	$12 \div 4 =$ _____	
7.	$3 \times 4 =$ _____	
8.	$4 \times 4 =$ _____	
9.	$5 \times 4 =$ _____	
10.	$12 \div 4 =$ _____	
11.	$16 \div 4 =$ _____	
12.	$20 \div 4 =$ _____	
13.	$6 \times 4 =$ _____	
14.	$7 \times 4 =$ _____	
15.	$8 \times 4 =$ _____	
16.	$24 \div 4 =$ _____	
17.	$28 \div 4 =$ _____	
18.	$32 \div 4 =$ _____	
19.	$9 \times 4 =$ _____	
20.	$10 \times 4 =$ _____	
21.	$36 \div 4 =$ _____	
22.	$40 \div 4 =$ _____	

23.	$4 \times$ _____ $= 8$	
24.	$4 \times$ _____ $= 16$	
25.	$8 \div 4 =$ _____	
26.	$16 \div 4 =$ _____	
27.	$4 \times$ _____ $= 24$	
28.	$4 \times$ _____ $= 32$	
29.	$24 \div 4 =$ _____	
30.	$32 \div 4 =$ _____	
31.	$4 \times$ _____ $= 4$	
32.	$4 \times$ _____ $= 0$	
33.	$4 \div 4 =$ _____	
34.	$8 \div 4 =$ _____	
35.	$4 \times$ _____ $= 12$	
36.	_____ $\times 4 = 20$	
37.	$4 \times$ _____ $= 28$	
38.	_____ $\times 4 = 36$	
39.	$4 \times$ _____ $= 40$	
40.	_____ $\div 4 = 3$	
41.	$20 \div$ _____ $= 4$	
42.	_____ $\div 4 = 7$	
43.	$36 \div$ _____ $= 4$	
44.	_____ $\div 4 = 10$	

Name

Complete the equation. Use the place value chart to help you.

1.

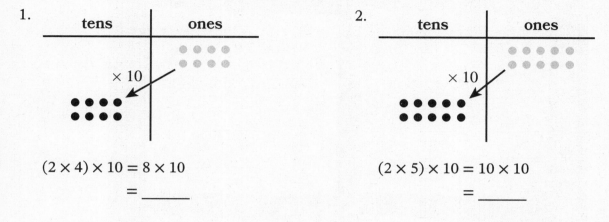

tens	ones

$\times 10$

$(2 \times 4) \times 10 = 8 \times 10$

$= \rule{2cm}{0.4pt}$

2.

tens	ones

$\times 10$

$(2 \times 5) \times 10 = 10 \times 10$

$= \rule{2cm}{0.4pt}$

3.

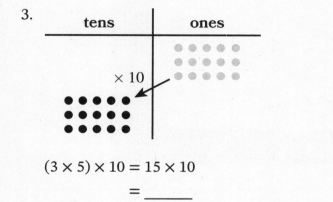

tens	ones

$\times 10$

$(3 \times 5) \times 10 = 15 \times 10$

$= \rule{2cm}{0.4pt}$

4.

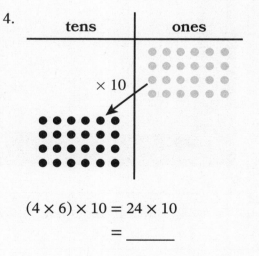

tens	ones

$\times 10$

$(4 \times 6) \times 10 = 24 \times 10$

$= \rule{2cm}{0.4pt}$

5. Place parentheses and fill in the blanks to find each related fact and product. The first two are started for you.

$2 \times 20 = 2 \times (2 \times 10)$

$= (2 \times 2) \times 10$

$= \underline{\qquad} \times 10$

$= \underline{\qquad}$

$2 \times 30 = 2 \times (3 \times 10)$

$= (2 \times 3) \times 10$

$= \underline{\qquad} \times 10$

$= \underline{\qquad}$

$3 \times 30 = 3 \times (3 \times 10)$

$= 3 \times 3 \times 10$

$= \underline{\qquad} \times 10$

$= \underline{\qquad}$

$2 \times 50 = 2 \times 5 \times 10$

$= 2 \times 5 \times 10$

$= \underline{\qquad} \times 10$

$= \underline{\qquad}$

6. Ivan finds 40×3 by thinking about how many tens are in 40. Explain Ivan's strategy.

_____

_____

_____

_____

Name _____

1. Place parentheses in the equations to find the related fact. Then complete the equation to find the product.

$$4 \times 20 = 4 \times 2 \times 10$$
$$= 4 \times 2 \times 10$$
$$= \underline{\hspace{1cm}} \times 10$$
$$= \underline{\hspace{1cm}}$$

2. Ray finds 8×30 by thinking about how many tens are in 30. Explain Ray's strategy.

22

Name

Use the Read–Draw–Write process to solve the problem. Use a letter to represent each unknown.

1. There are 8 rows of 10 carpet squares in each classroom. How many carpet squares are in 4 classrooms?

Use the Read–Draw–Write process to solve the problem. Use a letter to represent each unknown.

2. James wants to buy an art kit that costs $200. He saves $40 each month for 3 months. How much more money does James need to save to buy the art kit?

22

Name

Use the Read–Draw–Write process to solve each problem. Use a letter to represent each unknown.

1. There are 30 students in Miss Wong's class. Each student reads 5 books during March.

   a. What is the total number of books the students read in March?

   b. The students in Mr. Lopez's class read a total of 95 books in March.

   How many more books did Miss Wong's class read than Mr. Lopez's class?

2. Shen buys 8 sheets of stamps. Each sheet of stamps has 20 stamps.

   a. How many stamps does Shen buy?

   b. Shen already had 182 stamps. How many stamps does Shen have now?

3. There are 7 rows of 10 floor tiles in each classroom. What is the total number of floor tiles in 6 classrooms?

---

4. There are 60 minutes in 1 hour. Students are in school for 6 hours and 45 minutes.

   What is the total number of minutes students are in school?

5. Jayla recycles 48 cans and 32 bottles. She gets 5 cents for each can or bottle she recycles.

What is the total number of cents Jayla gets for recycling her cans and bottles?

✉ **22**

Name _____

Use the Read–Draw–Write process to solve the problem. Use a letter to represent each unknown.

Adam has 4 bags of marbles. Each bag has 30 green marbles and 40 yellow marbles.

What is the total number of marbles in all the bags?

Name _____

Use the pictures to help you complete the table.

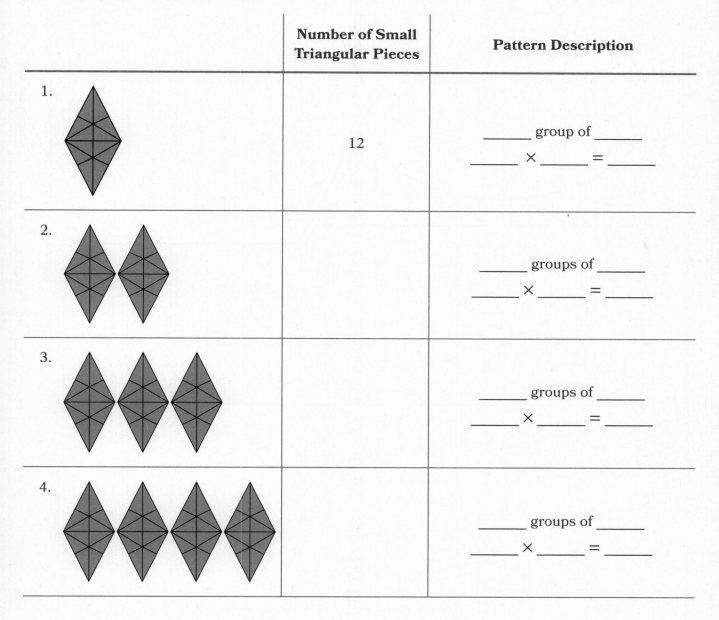

	Number of Small Triangular Pieces	Pattern Description
1.	12	_____ group of _____ _____ × _____ = _____
2.		_____ groups of _____ _____ × _____ = _____
3.		_____ groups of _____ _____ × _____ = _____
4.		_____ groups of _____ _____ × _____ = _____

5. Show three ways to find 12×8.

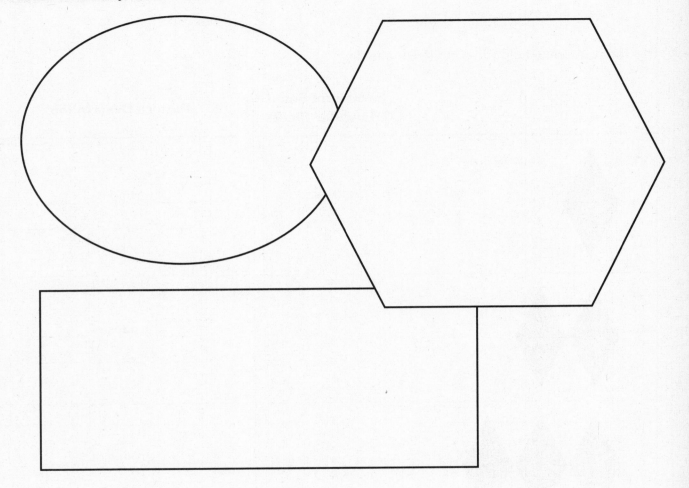

23

Name _____

1. Skip-count by 11.

___0___ , ___11___ , _____ , _____ , _____ , _____ , _____ , _____ , ___88___ , _____ , _____

Find the value of each unknown.

2. $2 \times 11 = m$

3. $r \times 11 = 44$

4. $f = 11 \times 6$

5. $j \times 7 = 77$

6. $11 \times 8 = a$

7. $99 = 11 \times h$

8. Ray knows that $9 \times 11 = 99$. He uses 99 to find 10×11.

a. Fill in the blanks to show Ray's strategy.

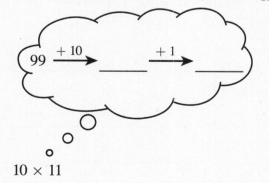

10×11

b. Complete the multiplication equation.

$10 \times 11 =$ _____

9. Skip-count by 12. Fill in the blanks to help you.

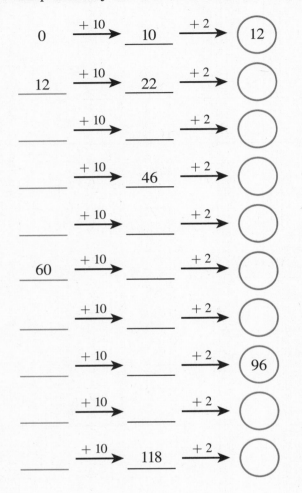

10. How does adding 10 and 2 more help you skip-count by 12?

_____

_____

_____

_____

Find the value of each unknown.

11. $3 \times 12 = c$

12. $60 = n \times 12$

13. $w = 12 \times 8$

14. $y \times 9 = 108$

15. Liz draws a tape diagram to represent 12×7. Fill in the blanks to find the total.

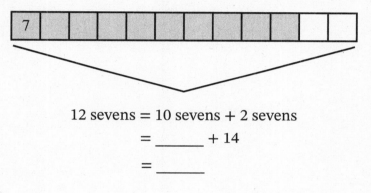

12 sevens = 10 sevens + 2 sevens

= _____ + 14

= _____

Use the Read–Draw–Write process to solve each problem.

16. 7 soccer teams have 11 players each. How many players are there?

17. How many eggs are in 6 dozen?

Name

Find each product. Show your work.

1. $9 \times 12 =$ _____

2. $11 \times 7 =$ _____

Collection 1

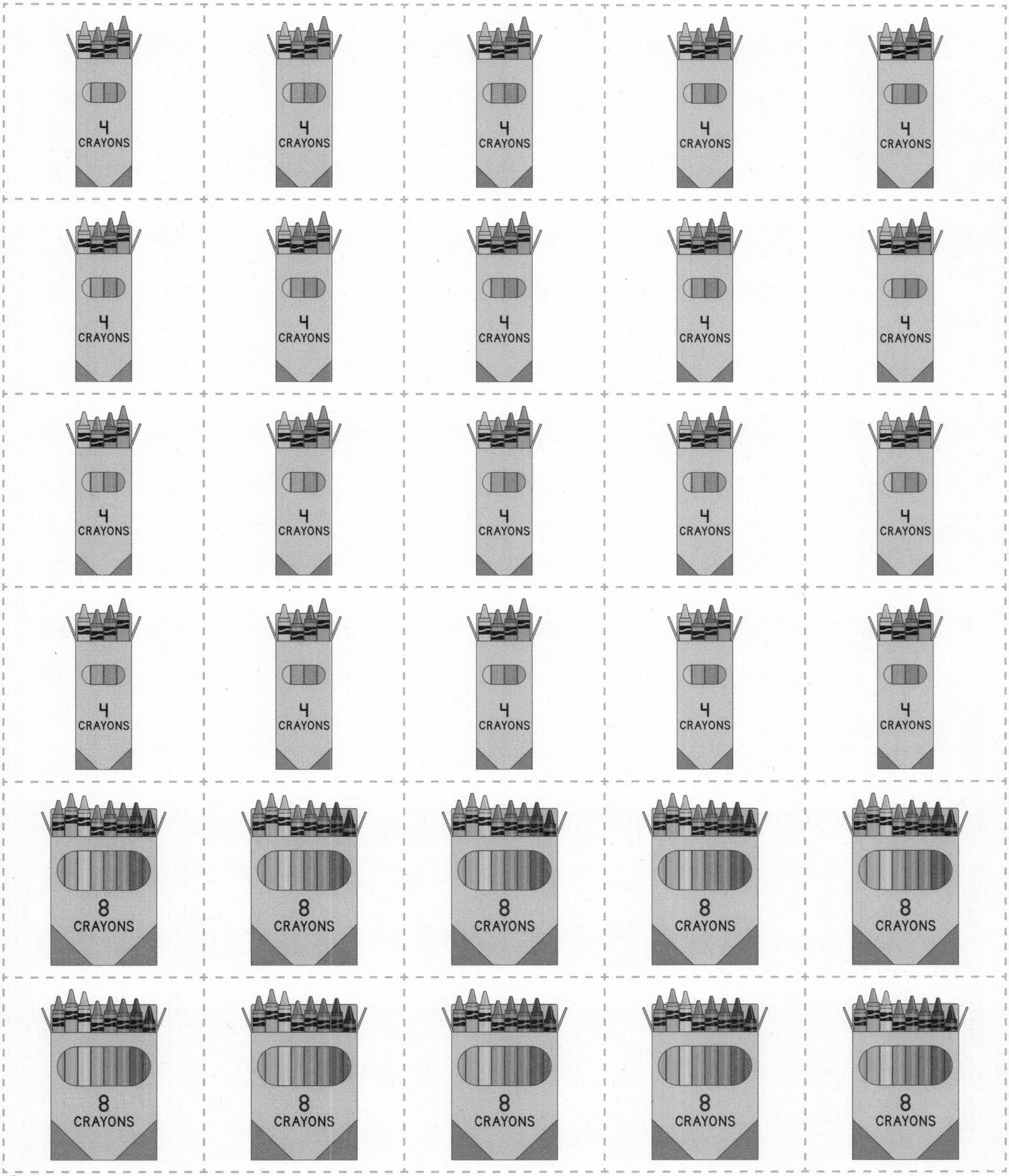

Collection 2

7 ERASERS   **7 ERASERS**   **7 ERASERS**   **7 ERASERS**   **7 ERASERS**

7 ERASERS   **7 ERASERS**   **7 ERASERS**   **7 ERASERS**   **7 ERASERS**

7 ERASERS   **7 ERASERS**   7 ERASERS   7 ERASERS   7 ERASERS

7 ERASERS   7 ERASERS   7 ERASERS   7 ERASERS   7 ERASERS

7 ERASERS   7 ERASERS

Collection 3

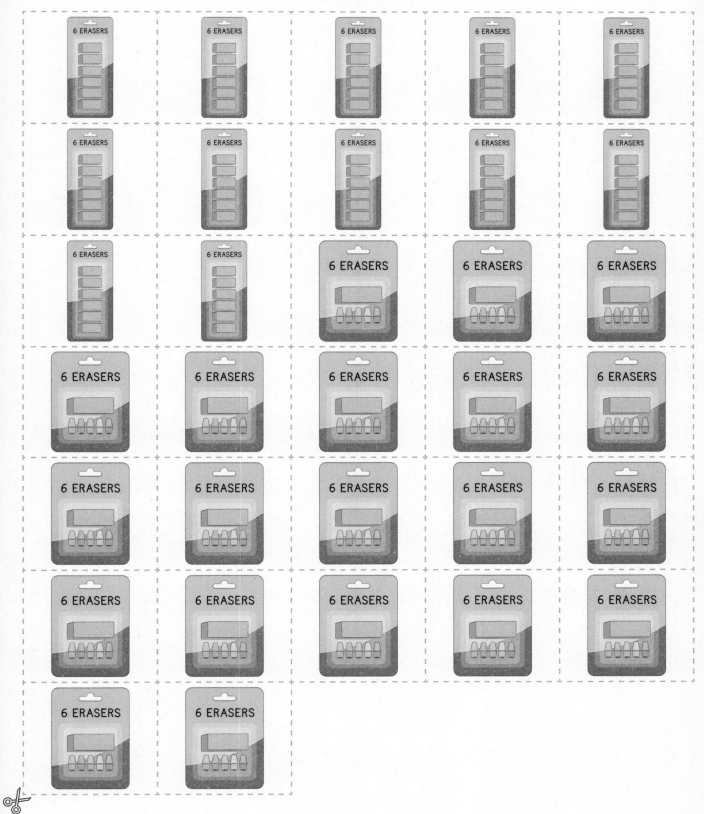

Collection 4

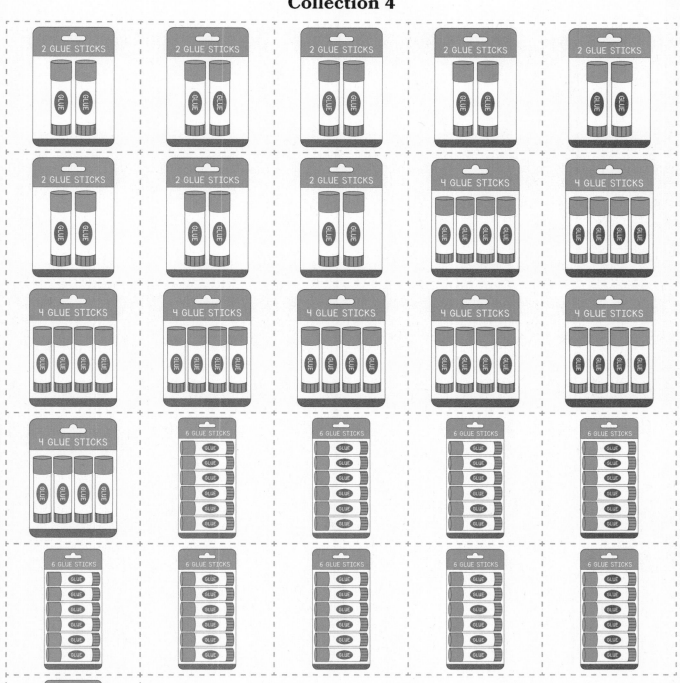

Collection 5

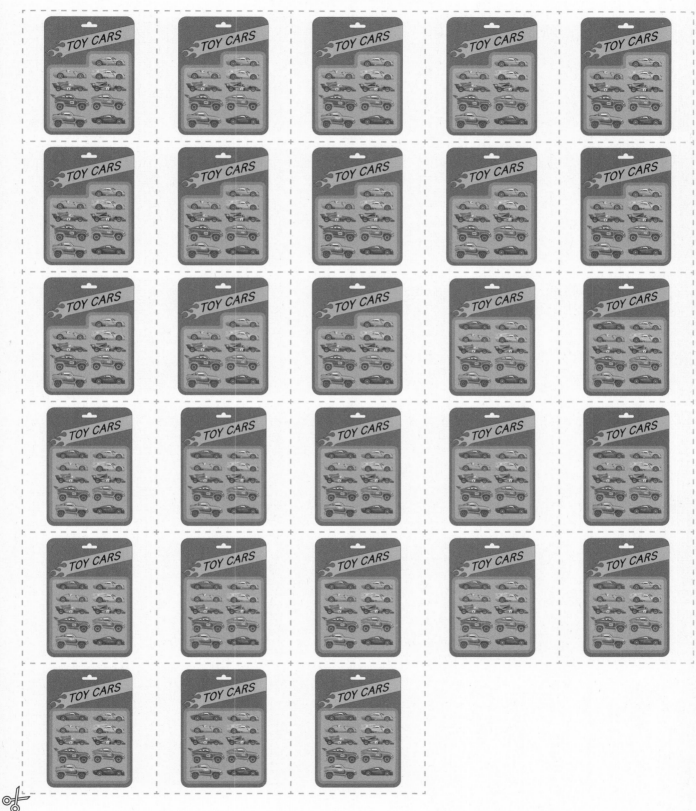

Collection 6: Mixed Collection

Collection 7: Mixed Collection

7 ERASERS

7 ERASERS

7 ERASERS

7 ERASERS

7 ERASERS

7 ERASERS

7 ERASERS

7 ERASERS

6 CRAYONS

6 CRAYONS

6 CRAYONS

6 CRAYONS

6 CRAYONS

6 CRAYONS

12 GLUE STICKS

12 GLUE STICKS

12 GLUE STICKS

12 GLUE STICKS

Name _____

1. Fill in the blanks so that each expression equals 6 × 12. Then choose one expression and use it to find 6 × 12.

6 × 12

6 × (_____ + 2)   6 × (6 + _____)

6 × (5 + _____)   6 × (5 + _____ + 2)

Name _____

For this counting collection, I am partners with _____.

We are counting _____.

We estimate there are about _____ of them.

This is how we organized and counted the collection:

We counted _____ altogether.

An equation that describes how we found the total is:

_____

Self-Reflection

Write one thing that worked well for you and your partner. Explain why it worked well.

_____

_____

_____

Write one challenge you had. How did you work through the challenge?

_____

_____

_____

_____

Name

1. What new strategies did you use to count?

2. How is the strategy you used to count today more efficient than the strategy you used to count the last time?

Name _____

Adam goes shopping for a party. He has $100 to buy supplies. The table shows the cost of each supply.

Supply	Cost
Pack of Small Plates	$4
Pack of Large Plates	$6
Pack of Napkins	$3
Balloons (Bag of 20)	$2
Tablecloth	$5

Use the Read–Draw–Write process to solve each problem.

1. Adam buys 5 packs of napkins. What is the total cost of the napkins?

2. Adam needs 75 balloons. He buys 4 bags of balloons.

   a. Did he buy enough balloons? How do you know?

   b. What is the total cost of the balloons?

3. Adam spends $28 on plates. Show two different ways that he could spend $28 on plates.

---

4. Does Adam have enough money left to buy 6 tablecloths? How do you know?

Name

Use the Read–Draw–Write process to solve each problem.

Food	Cost of Each
Pizza	$10
Fruit Tray	$32
Ice Cream	$4
Cupcakes (Pack of 6)	$9

Adam gets ready for a party. 48 people will be at the party.

1. Adam buys 8 pizzas. What is the total cost of the pizzas?

2. Adam buys 10 packs of cupcakes.

    a.   Does he buy enough cupcakes for everyone to have 1 cupcake? Explain how you know.

    b.   What is the total cost of the cupcakes?

    c.   Adam puts the cupcakes in rows of 5. How many rows of cupcakes are there?

3. Adam spends $48 on a fruit tray and ice cream. How many containers of ice cream does he buy?

4. What is the total cost of the food that Adam buys?

5. Adam sets up 6 tables. How many chairs should he put at each table so everyone has a chair?

6. There are 2 vases of flowers on each table. Each vase has 7 flowers. How many flowers are there in total?

7. Adam sets up 3 rows of 9 chairs for the children to watch a magic show.

   a. How many children will be at the party?

   b. How many adults will be at the party?

Name

Gabe shops for office supplies. The table shows the cost of each supply.

Supply	Cost
Pencils (Pack of 15)	$3
Pens (Pack of 20)	$4
Box of Staples	$5
Box of Paper Clips	$6
Stapler	$7

1. Gabe buys 8 packs of pens.

   a. What is the total cost of the pens?

   b. How many pens does Gabe buy?

2. Gabe needs 50 pencils. He buys 3 packs of pencils.

   Did he buy enough pencils? How do you know?

Credits

Great Minds® has made every effort to obtain permission for the reprinting of all copyrighted material. If any owner of copyrighted material is not acknowledged herein, please contact Great Minds for proper acknowledgment in all future editions and reprints of this module.

Cover, Paul Klee, (1879–1940), *Farbtafel "qu 1"* (Colour table "Qu 1"), 1930, 71. pastel on coloured paste on paper on cardboard, 37.3 x 46.8 cm. Kunstmuseum Basel, Kupferstichkabinett, Schenkung der Klee-Gesellschaft, Bern. © 2020 Artists Rights Society (ARS), New York.; pages 5, 217, 227, Modified, original image Viktorija Reuta/Shutterstock.com; pages 5, 217, Modified, original image pticelov/Shutterstock.com; pages 7, 9, 219, 221, 227, 229, Modified, original images Jo Ann Snover/Shutterstock.com and Feng Yu/Shutterstock.com; pages 7, 219, Modified, original image Jake Rennaker/Shutterstock.com; pages 9, 221, Modified, original image Feng Yu/Shutterstock.com; pages 11, 223, 227, 229, Modified, original image Kotema/Shutterstock.com; pages 13, 225, Modified, original image VectorShow/Shutterstock.com; page 117, iQoncept/Shutterstock.com; page 198, Modified, original images Valentin Agapov/Shutterstock.com and Gargantiopa/Shutterstock.com; All other images are the property of Great Minds.

For a complete list of credits, visit http://eurmath.link/media-credits.

Acknowledgments

Kelly Alsup, Lisa Babcock, Cathy Caldwell, Mary Christensen-Cooper, Cheri DeBusk, Jill Diniz, Melissa Elias, Janice Fan, Scott Farrar, Krysta Gibbs, Julie Grove, Karen Hall, Eddie Hampton, Tiffany Hill, Robert Hollister, Rachel Hylton, Travis Jones, Liz Krisher, Courtney Lowe, Bobbe Maier, Ben McCarty, Maureen McNamara Jones, Cristina Metcalf, Melissa Mink, Richard Monke, Bruce Myers, Marya Myers, Geoff Patterson, Victoria Peacock, Marlene Pineda, Elizabeth Re, Meri Robie-Craven, Jade Sanders, Deborah Schluben, Colleen Sheeron-Laurie, Jessica Sims, Theresa Streeter, Mary Swanson, James Tanton, Julia Tessler, Saffron VanGalder, Jackie Wolford, Jim Wright, Jill Zintsmaster

Trevor Barnes, Brianna Bemel, Adam Cardais, Christina Cooper, Natasha Curtis, Jessica Dahl, Brandon Dawley, Delsena Draper, Sandy Engelman, Tamara Estrada, Soudea Forbes, Jen Forbus, Reba Frederics, Liz Gabbard, Diana Ghazzawi, Lisa Giddens-White, Laurie Gonsoulin, Nathan Hall, Cassie Hart, Marcela Hernandez, Rachel Hirsh, Abbi Hoerst, Libby Howard, Amy Kanjuka, Ashley Kelley, Lisa King, Sarah Kopec, Drew Krepp, Crystal Love, Maya Márquez, Siena Mazero, Cindy Medici, Patricia Mickelberry, Ivonne Mercado, Sandra Mercado, Brian Methe, Mary-Lise Nazaire, Corinne Newbegin, Max Oosterbaan, Tamara Otto, Christine Palmtag, Andy Peterson, Lizette Porras, Karen Rollhauser, Neela Roy, Gina Schenck, Amy Schoon, Aaron Shields, Leigh Sterten, Mary Sudul, Lisa Sweeney, Samuel Weyand, Dave White, Charmaine Whitman, Nicole Williams, Glenda Wisenburn-Burke, Howard Yaffe

Talking Tool

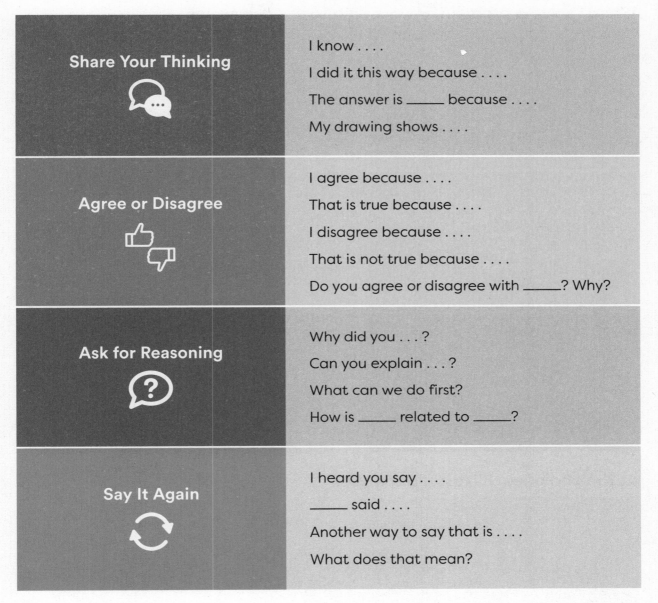

Share Your Thinking	I know . . . . I did it this way because . . . . The answer is _____ because . . . . My drawing shows . . . .
Agree or Disagree	I agree because . . . . That is true because . . . . I disagree because . . . . That is not true because . . . . Do you agree or disagree with _____? Why?
Ask for Reasoning	Why did you . . . ? Can you explain . . . ? What can we do first? How is _____ related to _____?
Say It Again	I heard you say . . . . _____ said . . . . Another way to say that is . . . . What does that mean?

Thinking Tool

When I solve a problem or work on a task, I ask myself

Before	Have I done something like this before?
	What strategy will I use?
	Do I need any tools?
During	Is my strategy working?
	Should I try something else?
	Does this make sense?
After	What worked well?
	What will I do differently next time?

At the end of each class, I ask myself

What did I learn?

What do I have a question about?